GENETIC DIVERSITY

DIVERSITY

And HUMAN

BEHAVIOR

GENETIC DIVERSITY

And HUMAN BEHAVIOR

J. N. *Spuhler,* editor

Routledge
Taylor & Francis Group

LONDON AND NEW YORK

First published 1967 by Transaction Publishers

Published 2017 by Routledge
2 Park Square, Milton Park, Abingdon, Oxon OX14 4RN
711 Third Avenue, New York, NY 10017, USA

Routledge is an imprint of the Taylor & Francis Group, an informa business

Library of Congress Catalog Number: 2008055077

Library of Congress Cataloging-in-Publication Data
Genetic diversity and human behavior / J.N. Spuhler.
     p. cm.
   Papers presented at a symposium sponsored by the Wenner-Gren Foundation for Anthropo-
     logical Research, and held at Burg Wartenstein, Austria, Sept. 16-28, 1964.
   ISBN 978-0-202-36308-0 (alk. paper)
     1. Behavior genetics--Congresses. 2. Human behavior--Congresses. 3. Human genet-
ics--Variation--Congresses. I. Spuhler, J. N. (James N.), ed. II. Wenner-Gren Foundation for Anthropological Research.

QH431.G396 2009
155.7--dc22

                                                         2008055077

ISBN 13: 978-0-202-36308-0 (pbk)

PREFACE

THESE PAPERS on human behavioral genetics were first presented in the Burg Wartenstein Symposium No. 27 entitled "Behavioral Consequences of Genetic Differences in Man," meeting September 16–28, 1964, in the European conference center of the Wenner-Gren Foundation for Anthropological Research, New York, at The Burg, near Gloggnitz, Austria. The conference was planned on the invitation of the late Dr. Paul Fejos by the members of the Committee on Genetics and Behavior, Social Science Research Council, New York, in collaboration with the foundation staff.

Although some anthropologists and psychologists had long been unusual among behavioral and social scientists in promulgating a program that included the possibility of a broad junction between genetics and the study of human behavior, no well-ordered body of knowledge that deserved to be called "human behavioral genetics" existed until well after World War II. The first *Annual Review of Psychology* article on "Behavior Genetics" was written in 1960, the same year Fuller and Thompson published the first textbook on the subject.

To be sure, a few important works on the subject (especially those using observations on twins) appeared after Galton's pioneer work, *Inquiries into Human Faculties* (1883), and before 1960: for example, Muller, "Mental Traits and Heredity" (1925); Newman, Freeman, and Holzinger, *Twins: A Study of Heredity and Environment* (1937); and Burt and Howard, "The Multifactorial Theory of Inheritance and Its Application to Intelligence" (1956). Yet most anthropologists were still impressed with the idea that nearly all cultural content was learned, shared, and transmitted by symbolic interchange largely independent of genotypic diversity within and between local groups—so much so that men to them became merely neutral containers to be filled and emptied without affecting the cultural content itself.

The period following World War II saw an acceleration of activity in the fields in and bordering on behavioral genetics. New results in neuroendocrinology closed the circuit showing the higher cortical centers could influence and be influenced by the hypothalamus, pituitary, thyroid, adrenals, and gonads. Genetic diversity in the function of these organs had obvious consequences for social and cultural behavior. The failure of some early and long-reinforced attempts at conditioning by students of comparative animal behavior showed species-specific innate behavior could not be ignored in a theory that attempted to combine psychology and anthropology. These are examples of some of the developments that led to the organization of the symposium reported here.

Illness and an unavoidable conflict in time prevented three persons—from England, France, and the Netherlands—from attending the conference; their absence explains the lack of papers in this book on implications of chromosomal abnormalities for mental defect and on biochemical aspects of behavioral conditions, such as phenylketonuria and Huntington's chorea, controlled by certain major genes. Since excellent reviews of new developments in these fields are available, I have not attempted to obtain substitute papers written especially for this volume. I do think it is important, however, to mention that, had these three individuals from Europe been able to attend the conference, the balance between nations of residence would have been more in line with that desired by the planners of the symposium.

Four participants in the Wartenstein conference did not furnish manuscripts for this volume. Jan A. Böök, University of Uppsala, presented materials on genetics and psychopathology; his findings may be consulted in recent papers from the Institute of Medical Genetics in Uppsala.

L. L. Cavalli-Sforza, University of Pavia, reported his investigations on the determinants of inbreeding in Italian village and city populations and on his progress (with A. W. F. Edwards and others) in developing a numerical phyletic taxonomy for man based on the statistical notion of minimum path and the number of gene substitutions. This work is published in the *Proceedings* of the eleventh International Congress of Genetics, The Hague, 1963, and in *Cold Spring Harbor Symposia on Quantitative Biology*, Long Island, 1964.

Irvin DeVore, Harvard University, discussed the implications for the study of man which comes from systematic observations on the behavior of non-human primates in their natural setting. This material is published in a volume edited by DeVore, *Primate Social Behavior* (Holt, Rinehart, and Winston, 1965).

Louis Guttman, Israel Institute of Applied Social Research, Jerusalem, reviewed some aspects of multivariate structural analysis with much value to the study of human behavioral genetics; their application is exemplified in the work reported in this book by Ruth Guttman.

Limitations of space prevented publication of a monograph-length work on Morphological Variation and Human Behavior prepared for the conference by Friedrich Keiter. An enlarged version of the paper presented here is scheduled for publication under the title *Die naturalische Themenordung des Menschenlebens*, by Reinhardt, Munich, in 1966.

The papers by David Rodgers and Delbert Thiessen were made available for publication here at my request.

The title of the summary paper by Ernst Caspari repeats the original name of the Wartenstein symposium. His paper is a summary of the actual discussions at the symposium and therefore mentions a number of points made in those discussions but not presented in the papers reproduced here. And, naturally, he could not refer to the material in those papers that were prepared or modified after the conference.

For all of us who took part in the conference it is a real pleasure to thank Lita

Osmundsen, Director of Research, Wenner-Gren Foundation, and her unusually able staff for giving us a comfortable and enjoyable stay at Burg Wartenstein, plus excursions to Burgenland and the opera in Vienna. The staff of the foundation make "The Castle" one of the more attractive conference centers in Europe—*Vielen Dank für Speis' und Trank.*

J. N. SPUHLER

CONTENTS

GENETIC DIVERSITY AND HUMAN BEHAVIOR

ON TYPES, GENOTYPES, AND
THE GENETIC DIVERSITY IN POPULATIONS

THEODOSIUS DOBZHANSKY

GENETIC DIFFERENCES in man, like those between individuals of any animal or plant species and those between species, are all products of the evolutionary development of the living world. These differences, with their behavioral consequences, can only be understood in the light of evolution, *sub specie evolutionis*. Our understanding of evolution, however, is itself evolving. The Darwin-Wallace theory of evolution appeared more than a century ago. The development of the evolutionary thought has gone through several stages. It was necessary, first of all, to demonstrate beyond reasonable doubt that evolution had, in fact, taken place. Roughly the first four decades after Darwin, until about 1900, were dominated by studies in comparative anatomy, embryology, systematics, zoogeography, phytogeography, and paleontology, designed to discover and examine the evidences of evolution. During the same period, the first sketchy attempts were made to trace the paths that the evolution of the living world actually followed. The evidences of evolution are now so well known that they are part of the basic college and even high school courses of biology. On the contrary, the phylogenies of the animal and plant kingdoms, and particularly the phylogeny of man, are far from completely known, and they are actively studied today.

The study of the causes that bring evolution about came to the forefront of attention at the turn of the century. The first three decades of the current century were mainly a period of data-gathering. This was the time of the rapid growth of basic genetics, and genetics was more often than not studied for its own sake, rather than because of its bearing on evolutionary problems. Nevertheless, although a causal analysis of evolution grounded in genetics may well be wrong, one flying in the face of the established principles of genetics cannot possibly be right. The next twenty years or so, roughly from 1930 to 1950, saw the gradual formulation of the modern biological, also called the synthetic, theory of evolution. Adumbrated by Chetverikov in 1926, the basic principles of the biological theory were arrived at almost simultaneously by Fisher, Haldane, and Wright around 1930. For the first time in the history of biology a biological theory was deduced mathematically from, ultimately, a single fundamental premise—Mendel's law of segregation. The theory was soon shown to possess a great explanatory power in biology as a whole. Within two decades Mayr, Simpson,

Rensch, Schmalhausen, Stebbins, Darlington, White, and others demonstrated that the theory makes sense of experimental genetics, paleontology, zoological and botanical systematics, cytology, comparative morphology, and embryology. More recently, the theory has been applied successfully in anthropology, ecology, physiology, and biochemistry. It has been strengthened by the spectacular developments in molecular genetics.

Historical perspectives tend to become blurred when one approaches the present. Where are we now and where are we going? Has the theory of evolution congealed into a dogma? Far from that, evolutionary thought is now in a period of rapid development and change. I agree with Mayr (1963) that "the replacement of typological thinking by population thinking is perhaps the greatest conceptual revolution that has taken place in biology." My only cavil is that Mayr's "has taken place" is overoptimistic; I do believe, however, that this conceptual revolution is under way.

The differences between typological and populational thinking are often as subtle as they may be profound. Many biologists are completely unaware of these differences, and yet Mayr is right that "virtually every major controversy in the field of evolution has been between a typologist and a populationist." To a typologist, the diversity of living individuals is illusory. Individuality is an accidental deviation from the one real, constant, and unchangeable type of the species to which the individual belongs. It is the basic species type that a scientist should endeavor to discover and to understand. To a populationist, concretely existing individuals, not types, are the biological entities. Individuals may even be the ontological entities. The differences between individuals must be investigated, not ignored, and explained, not explained away.

Averages are statistical abstractions, and types are at best convenient models or devices that an observer utilizes to facilitate the description and communication of the data he has gathered. The use of "types" for this purpose is, of course, a legitimate procedure. A textbook of zoology can hardly avoid using schemes, or "body plans," of a worm, an insect, a vertebrate animal, and so forth. Such schemes or "types" are abstractions that represent the common features of a group of organisms and omit the characteristics by which they differ. In a sense, any classification involves some kind of abstraction, and in some fields of study (for example, archeology and linguistics), the word "typology" is used in a way that makes it a near synonym of what in biology is called "classification." Classification and schematization, however, cease to be legitimate, and become specimens of typological thinking, when the classifier or schematizer forgets that the animals, plants, and people exist before and independently of any schemes. Typological thinking is what in philosophy is called "reification" of abstractions. Populational thinking, especially when applied to man, becomes closely related to individualism. It could also be called existential thinking, if it were not that the word "existentialism" in recent years has been used in so many different senses that it has become utterly ambiguous.

PHILOSOPHICAL ANTECEDENTS

The roots of both the typological and the populational approaches go deep in the history of human thought. As Bronowski (1956) beautifully put it, "Science is nothing else than the search to discover unity in the wild variety of nature—or more exactly, in the variety of our experience. Poetry, painting, the arts are in the same search, in Coleridge's phrase, for unity in variety. Each in its own way looks for likeness under the variety of human experience." The oldest and most important instrument whereby the runaway variety is bridled is in symbolic language. Words like "man," "cat," or "dog" refer not to particular persons or animals but to representatives of mankind, cat-kind, and dog-kind. Human language is a spontaneous typologist.

The diversity of persons, cats, and dogs is nevertheless inevasible. Moreover, individual persons and animals change and grow old. It is tempting to declare the diversity and change to be false fronts or spurious appearances. Parmenides did so about 24 centuries ago. A similar result was achieved by Plato more gently, by his famous theory of ideas. God has created the eternal, unchangeable, and inconceivably beautiful prototypes, or ideas, of Man, of Horse, and even of such mundane and inanimate objects as Bed and Table. Individual persons, horses, beds, and tables are only pale shadows of their respective ideas. Acquisition of wisdom is a way to catch a glimpse of the ideas that one formerly saw only as shadows.

Platonic philosophy, of course greatly modified in form but still true to its spirit, has survived the centuries and millennia. It is a living influence today. Aristotle did not like the Platonic ideas, but assumed one cosmic idea which manifests itself in the visible world. The realization of the ideal form is the purpose of nature, most clearly observable in the living world. All animals are variants of a single architectonic plan. For more than two millennia, until the time of Darwin, the organic world was viewed as either Plato or Aristotle saw it. The living, and even the nonliving, bodies formed a "Great Chain of Being" (Lovejoy, 1960), ranging from a lesser to greater perfection. Leibniz in the seventeenth and Bonnet in the eighteenth centuries felt satisfied that the Chain is single and uninterrupted. Bonnet saw it starting with fire and "finer matters," and extending through air, water, minerals, corals, truffles, plants, sea anemones, birds, ostriches, bats, quadrupeds, monkeys, and so to man.

To Lamarck and to Geoffroy St. Hilaire, but not to their predecessors and contemporaries, the Great Chain implied evolution. Has the world actually evolved through the stages corresponding to the links of the Chain? In the famous debate between St. Hilaire and Cuvier in 1830, the issue of evolution was stultified by being mixed up with the Great Chain. Cuvier was, of course, absolutely right that there is no discernible single plan of body structure common to all animals. The hypothesis that living organisms are manifestations of a limited number of basic types or ideas of structure nevertheless proved to be very useful. It inspired studies on comparative anatomy and classification from Belon, through Buffon, Daubenton, Goethe, Oken, and the school of German *Natur-philosophie*, to St. Hilaire, Karl von Baer, and Richard Owen.

There is no doubt that the data collected by these investigators furnished a base for the doctrine of evolution. It is, however, questionable whether they can validly be regarded as pioneer evolutionists. Goethe's *Urpflanze* was not claimed to have actually grown and flowered anywhere in particular; still more important, the existing plants were not claimed to have the *Urpflanze* as their actual ancestor; the *Urpflanze* was merely a blueprint according to which the "creative power" (*schopfende Gewalt*) designed the existing plants. That Baer and Owen strenuously opposed Darwin should come as no surprise. As late as 1876, Baer insisted that "one must clearly distinguish the ideal from the genetic or genealogical kinship," and, while favoring the former, denied the latter. Owen was even more explicit; his morphological archetypes were expressions of Plato's eternal ideas (see Zimmermann, 1953, and Kanaev, 1963, for anthologies and discussions of the history of evolutionary thought).

The roots of the populational thought are less easily traced than those of typology. The former arises primarily in connection with observations on human beings, rather than with studies on animals and plants. Human individuality is stressed in most religions, and is one of the tenets of the Christian world outlook. The situation is complicated because that outlook has also a typological component—the doctrine of Original Sin, by which human nature is held to be ineluctably debauched. A geneticist is tempted to imagine that the fruit that Adam and Eve so imprudently ate contained some powerful mutagen, which induced a heavy genetic load in all their descendants. It is, however, an individual person who commits sins; we possess free wills, and are accountable for our actions. Pelagius stated this most clearly (too clearly, it would appear, since he was declared a heretic):

Everything good and everything evil, in respect of which we are either worthy of praise or of blame, is done by us, not born with us. We are not born in our full development, but with a capacity for good and evil; we are begotten as well without virtue as without vice (quoted after Dunham, 1964).

The formal philosophical codification of human individuality came only in 1785, in Kant's doctrine that every human being is an end in himself. This doctrine is, however, implicit in the belief that every man has certain inalienable rights; this belief became popular during the Age of Enlightenment, and it serves as the foundation of the theory of democracy. In general, the typological view of man goes most easily with conservative and the view that every man is an end in himself with liberal political persuasions.

But man's efforts to know himself are often frustrated by his remarkable capacity to deceive himself. Quite inconsistently, the conservatives also favor the view that different people are genetically different, while the liberals prefer the belief that man is at birth a *tabula rasa*, a blank slate, and that all differences are the products of upbringing and education.

EVOLUTION VS. TYPOLOGY

Darwin wrote that his great work, *The Origin of Species*, was "one long argument." The heart of this long argument was "that species are only strongly marked and permanent varieties, and that each species first existed as a variety." Varieties, hence, are not merely diverse manifestations of the same archetype, of the Platonic idea of their species; they are themselves incipient species, and the existing species are aggregations of varieties. Now, Darwin used the word "variety" in a sense so inclusive that it became ambiguous. His "varieties" ranged from what we would call subspecies or races—that is, genetically distinct populations occupying definite geographic territories to the exclusion of other races of the same species—to breeds of domestic animals and cultivated plants, and down to polymorphs and aberrant individuals found within an interbreeding population. This ambiguity proved so confusing that "variety" is no longer used as a technical term. In Darwin's "long argument," however, his use of the word was not unreasonable; what Darwin set out to demonstrate was that there existed within a species genetic raw materials from which natural selection could compound new species.

Platonic types exist in some realm above the world of human senses; they are intrinsically static and unchangeable. The theory of evolution has shattered the typological statics, and replaced it with a populational dynamism. The living world is changeable and capable of transformations; some of the transformations may represent progress and others degradation. Moreover, the changeability is not a false front or an illusion, as the followers of Parmenides and Plato thought; it may produce real novelties. An individual changes with age, but this change follows, within limits, a fixed pattern. A population may change in a variety of directions; it owes its plasticity to the genetic diversity within it. Genetic differences in man signify that the human species has changed, and may change in the future.

Typological thinking is un-evolutionary or anti-evolutionary. With the general acceptance in biology of evolutionism, typological thinking should have been, it would seem, replaced by populational thinking. This has not happened everywhere, because established habits of thought are often refractory to change. Not all biologists are aware of the profound incompatibility of evolution with typology. As pointed out above, biological classification, and perhaps any classification, goes easily with typological thinking. If one strives to delineate species or races, the presence of intermediate forms is a nuisance that one is tempted to shrug off whenever possible. Borderline cases between race and species are a godsend to an evolutionist but an annoyance to a museum man, whose business is to write a label for every specimen in his collection. Although the usefulness of classification is unquestionable, a maximal neatness of the taxonomic pigeonholes is not the ultimate aim of systematics. A modern systematist classifies the living world in order to understand it, and he is forced to pay more than lip service to evolution.

To illustrate the subtle but pervasive differences between the typological and the populational approaches, two examples may be considered briefly. The current civil rights movement in the United States has prompted the racists to devise a curious quasi-scientific justification of their social and political biases. It is alleged that, compared with whites, Negroes have a smaller average brain size and lower average scores on various intelligence and achievement tests. That these differences are questionable, and if real may be of environmental origin, need not concern us here. The point is that not even racists deny that all these measurements vary among whites as well as among Negroes. The variation ranges overlap so broadly that the large brains and the high scores in Negroes are decidedly above the white averages, and the small brains and the low scores in whites are much below the Negro averages.

Now, the typological reasoning goes about as follows. All Negroes are but manifestations of the same Negro archetype, and all whites of the same white archetype; no matter what his brain size or his intelligence score may be, an individual is either a Negro or a white, and should be treated accordingly. Some racists go so far as to claim, apparently in all sincerity, that it is unkind to Negroes to treat them as though they were white. The populational approach is simply that an individual is himself, not a pale reflection of some archetype. Every person has a genotype and a life history different from any other person, be that person a member of his family, clan, race, or mankind. Beyond the universal rights of all human beings (which may be a typological notion!), a person ought to be evaluated on his own merits.

One aspect of the controversy started by the recent book of Coon (1962) may be considered as our second example. According to this author, the species *Homo erectus* consisted of five subspecies, each of which was transformed into a subspecies of *Homo sapiens*, the transformation taking place in different places and at different times. Just what can this claim possibly mean? In a symposium held at Burg Wartenstein in 1962, and in a book published before Coon's (Dobzhansky, 1962, 1963), I argued that, at least from mid-Pleistocene on, there was but a single hominid species; that anagenesis (transformation) has been more important in the evolution of the human stock than cladogenesis (splitting); and that the species *Homo erectus* gradually evolved into *Homo sapiens*. Coon believes, however, that *sapiens* is an evolutionary "grade," which can be achieved independently, and earlier by one race than by another.

A species is an inclusive Mendelian population, a reproductive community, not a type or a grade. Becoming a new species is not instantaneous, like getting a new degree or being admitted into an exclusive club. The population of a species gradually changes its genetic composition, sometimes more rapidly and at other times more slowly. Eventually the accumulated genetic alterations become so clearly expressed in the fossil remains that a paleontologist is forced to apply a new label, in other words, a new species name. A typologist is then prone to assume that an individual to which a certain species label is applied is always more like any other individual that goes under the same species label than it is like an individual of any other species. This may be true only of species living contemporaneously, and only in a sense that an individual does not as a rule belong to more than one biological reproductive community. Even so, with respect to some genes an individual may resemble a representative of another species more than he resembles his sibling. For example, some humans and some chimpanzees have the A blood type and others have the O blood type. When an ancestral species is transformed in time into a descendant species, the criterion of reproductive community is no longer applicable. One obviously cannot tell whether individuals who lived many generations apart could or could not interbreed. They most certainly did not do so. An individual or a population of *Homo erectus* at a certain time level may then be more like one called *Homo sapiens* on the next higher level, than a more ancient population labelled *Homo erectus*. They were living populations, not typological "grades."

THE *TABULA RASA* THEORY

Racism is a form of typological misjudgment which assumes that an individual is a manifestation of a racial type. Mankind consists of several races or subspecies; the differences between the races are important, the differences among individuals within a race are trivial. However, another form of typology recognizes but a single type—that of the species, *Homo sapiens*. The classical statement of this view is John Locke's *tabula rasa* theory. This is frequently misrepresented as claiming that all people are similar at birth. Now, Locke was one of the most levelheaded among philosophers, and he did not claim anything so rash as that all infants are born alike. What he maintained is, rather, that there are no inborn "ideas." In his words,

Let us then suppose the mind to be, as we say, white paper, void of all characters, without any ideas; how comes it to be furnished? . . . All that are born into the world being surrounded with bodies that perpetually and diversely affect them, variety of ideas, whether care be taken about it or not, are imprinted on the minds of children.

The *tabula rasa* theory has had a great and abiding influence on the climate of "ideas" in which the civilized part of mankind lives. It has been adopted by the thinkers of the Age of Enlightenment, from Voltaire and Helvetius, Rousseau and Condorcet, to Bentham and Jefferson. It has been, and continues to be, an important ingredient in the philosophies of democracy. Its attractiveness lies in that it seems to uphold an optimistic view of human nature. A lot of people behave stupidly, wretchedly, and viciously. This is, we are assured, because they were badly brought up, and spoiled by corrupting influences of badly organized society. A better-organized society, better care, guidance, and education will, however, make everybody behave in accordance with the natural goodness and reasonableness that are the normal and universal endowment of every representative of the human species. Genetic differences among people have no behavioral consequences, at least not in a democratically organized society. I believe that

it can be demonstrated that, on the contrary, the assumption of an innate uniformity of all human beings makes nonsense of democracy. Though this problem
lies outside the frame of reference of the present discussion, I shall make some
brief comments about it below.

It is no exaggeration to say that different variants of the *tabula rasa* theory are
entertained, explicitly or implicitly, by most social scientists, and by some influential schools of psychology, especially those with psychoanalytical learnings.
Ruth Munroe lists as "the fourth basic concept accepted by all psychoanalytic
schools" that

Early childhood is the time when the malleable, adaptable, flexible human psyche takes
on the essential direction it will pursue. Because the human infant is so unformed by
nature, because these experiences are the first, the events of infancy have a psychological importance very naturally overlooked by the adult philosopher in his study
or the adult scientist among his instruments.

An idea carried to extremes may suddenly turn into its opposite. So it has happened with the *tabula rasa* concept. We have seen that some early Christian philosophers, particularly Pelagius, used an idea very similar to the *tabula rasa* notion
to affirm human individuality, in opposition to those who maintained that all
human natures are equally depraved by Adam's original sin. The same idea in
its modern guise makes the human nature uniform, and reduces individuality to
the status of a veneer applied by the infant-rearing practices and by circumstances
of a person's biography.

The trouble is that the *tabula rasa* theory starts with a valid premise but goes
on to draw an erroneous conclusion. The valid premise is that the behavioral development of *Homo sapiens* is remarkably malleable by external circumstances.
Man's paramount adaptive trait is his educability. In the broad sense in which it
is used here, *educability* means that man is able to adjust his behavior to circumstances in the light of experience. The genotype of the human species has been
so formed in the process of evolution that the educability is a universal property
of all nonpathological individuals. This universality, or near-universality, is a product of natural selection. It has made possible the human cultural development.
In all cultures, primitive and advanced, the vital ability is, and always was, to
be able to learn whatever is necessary to become a competent member of the
group or society in which the individual happens to be born or is placed by
circumstances.

Environmental plasticity, however, is not incompatible with genetic diversity.
Genetic differences do have behavioral consequences. To a geneticist this consideration is trivial, but he is constantly forced to remind his colleagues, especially those in the social sciences. It cannot be stressed too often that what is
inherited is not this or that "trait" or "character" but the way the development
of the organism responds to its environment. This is particularly important when
what is being considered is the genetic determination of the diversity in behavioral traits. The sex cells obviously cannot transmit "behavior"; what they do

transmit are genes, or, if you wish, chain molecules of deoxyribonucleic acids, which determine the pattern of the development of the organism, including the pattern of its behavioral development. The key word in the preceding sentence is "determine," and it must be understood correctly.

Genes determine the pattern of the development in the sense that, given a certain sequence of environmental influences, the development follows a certain path. The carriers of other genetic endowments in the same environmental sequence might well develop differently. But the development of the carrier of a given genotype might also follow different paths in different environments. The observed diversity of individual phenotypes would be more limited than it actually is if all people were as similar genetically as identical twins, but the environments were as manifold as they actually are; the diversity would be also more limited than it is, given the existing genetic variability but a uniform environment. To put it in another way, the observed phenotypic variance has both genetic and environmental components.

SOME PARAMETERS OF GENETIC DIVERSITY

Individuals of which a species is composed may be regarded as more or less faithful or defective incarnations of the species archetype. This is the typological approach. In modern biology it usually takes a disguise. For example, it is postulated that there exists somewhere, or can be obtained by breeding, the optimal genotype (= archetype), composed entirely of genes adaptively most favorable in a given environment, or even in all environments in which the species can live. Individuals observed in reality deviate more or less noticeably from the postulated possessors of the optimal genotype; they carry "genetic loads," which, in theory, can be measured by their deviations from the optimum genotype. The populational approach envisages a biological species as a Mendelian population composed of carriers of diverse genotypes. No genotype can really be "optimal," because the population faces a great variety of environments, inconstant in space and in time and with different adaptive requirements. How great is the genetic diversity within a given species is at best only sketchily known. In particular, the genetic diversity in the human species is explored quite inadequately. The matter needs more investigation, research, and analysis. This is one of the pivotal problems in the present book, and what I am trying to do in this introductory discussion is to place it in its proper perspective.

According to Ford's (1940) original definition, an interbreeding population is said to be genetically polymorphic if it consists of two or more clearly distinct and genetically different kinds of individuals. Some species and populations are obviously highly polymorphic, others less so, and still others seem to be monomorphic. The matter is, however, beset with complications. Our ability to recognize polymorphisms depends upon the techniques used. Before 1900 there was little reason to suppose that human populations are polymorphic for blood groups, because techniques were not available for diagnosing them. New biochemical polymorphisms are now being discovered in man almost annually, and the end is not yet in sight.

The immense field of polygenic variability is explicitly excluded by Ford's definition of polymorphisms, because polygenes, also by definition, do not produce discontinuous and phenotypically discrete variants. Yet at the biochemical level there is presumably no difference between the major genes and the polygenes. Some geneticists are loath to venture from the solid ground of discrete gene effects to the apparent morass of polygenes. The classical methods of genetics, recording the Mendelian segregation ratios in the progenies of hybrids, are not easily applicable to polygenic inheritance. (See, however, the paper by Thoday in the present volume.) Biometrical methods, selection experiments, and heritability determinations have to be used. Some geneticists have even recommended that a kind of moratorium be declared on studies of polygenic inheritance, until such time as it can be reduced to molecular terms and then presumably handled by the classical methods. I do not think this recommendation stands a chance of being adopted, for the simple reason that a majority of the differences between normal, nonpathological humans happen to be manifestations of polygenic differences. This field of study is too important to be relegated to an indefinite future time. Most relevant to our discussion, the nonpathological genetic variations in behavior traits in man seem to be mostly under polygenic control.

Compared to the number of persons who ever lived, the diversity of potentially possible genotypes in man is so vast that no more than a minute fraction of this diversity could be realized. The number of the genes in a human sex cell is not known. It could hardly be less than 10,000, which is the estimate, also not too reliable, usually given for the number of genes in *Drosophila*. With only two variant alleles per gene, this makes $2^{10,000}$ potentially possible genetically different gametes, and $3^{10,000}$ potentially possible zygotes. Many, or most, gene loci may be capable of giving more than two alleles per locus. Although not all potentially possible combinations have equal probabilities of actually being formed, the possible diversity is practically infinite. With every person (identical twins excepted) being genotypically different from every other, an immense majority of potentially possible genotypes will never be formed in reality.

The existing genotypes are, obviously, not just a random sample of the potentially possible ones. Even assuming that the mutation rates in some genes are as low as one per billion per generation, most genes will have actually mutated somewhere in the human species. However, the mutant genes will not necessarily persist in the populations. The dominant lethal mutations arising in a given generation will be extinguished in that same generation. Some of the spontaneous abortions, stillbirths, and neonatal deaths represent eliminations of the dominant, or semidominant, lethal mutant genes. A dominant mutation that would make its carrier unable to reproduce would not technically be called a lethal, but it will likewise be eliminated within a generation after its origin. This may be of interest to students of behavioral abnormalities which interfere with the reproduction of their carriers. Hutchinson (1959) called such abnormalities paraphilias; whether any of them are expressions of dominant mutant genes remains to be discovered.

Complete lethality, absolute sterility, and complete dominance are limiting cases. Semilethal and subvital genotypes, various degrees of lowered reproductive capacity, and semidominant and incompletely recessive genetic conditions are much more numerous in reality. Mutational changes giving rise to such conditions are of great interest and importance. The mutant genes persist in populations for an average number of generations, which is greater or smaller depending on the degree of the reduction of viability, reproductive capacity, and dominance. Completely recessive mutants, which according to some geneticists are very rare, will, of course, persist longer than incompletely recessive ones of the same degree of harmfulness. Simple mathematical formulas have been worked out relating these variables. It is not necessary for us to consider them here; what is important is that all living populations carry genetic loads consisting of more or less drastically deleterious mutants not yet eliminated by natural selection. The part of the genetic load maintained by the mutation pressure against selection is termed the mutational load.

The genetic variants that compose the mutational load are important components of the variability we observe in populations. This is because the genetic variants that cause slight or minute decreases of the adaptive values of their carriers, as measured by their reproductive performance, may be retained for quite considerable numbers of generations. Suppose that having a nose of an esthetically unappealing shape reduces the reproductive value of its carrier very slightly, say by one-thousandth, 0.001. If this nose shape is due to a single dominant gene arising by mutation once in 100,000 gametes (0.00001), then about 2 per cent of the population will have this nose shape. If a single recessive gene is responsible, then 10 per cent of the gametes will carry the gene but only 1 per cent of the population will have unattractive noses.

It is evidently difficult to obtain evidence for or against the hypothesis that the morphological, physiological, and behavioral variants found in human or other populations confer on their possessors adaptive advantages or disadvantages of the order of, say, 10 per cent and lower. This difficulty explains the persistence in population genetics of conflicting theories of population dynamics and particularly of genetic loads. This matter is one of crucial importance for understanding the nature of the genetic variation in mankind and in other biological species, and I shall discuss it as concisely and as objectively as I can.

The "classical" theory assumes that most of the unfixed, variable, genes are represented in populations by two or more alleles, one of which is normal, typical, and adaptively superior to the others; these latter make their carriers, if ever so slightly, inferior in fitness. Heterotic alleles, which make the heterozygotes superior to the homozygotes, are assumed to be rare. A genotype homozygous for all the normal alleles would evidently have the optimal fitness. The process of mutation, however, keeps injecting into the gene pool of the population deleterious mutant alleles; the deleterious alleles as yet unelimitated by the normalizing natural selection form the genetic load of the population; and the genetic load consequently is a mutational load. Ingenious statistical methods have actually been devised to measure the deviation of the genotypes from this theoretical optimal genotype.

The "balance" theory regards the optimal genotype a typological fiction. The populations of sexually reproducing and outbreeding species are arrays of diverse genotypes for several reasons other than the mutation process's interfering with the achievement of the optimal genotype. In the first place, it is unrealistic to imagine that any population exists in an absolutely uniform environment; since a population has to face a variety of environments, some genotypes highly fit in an environment A may be inferior in B or in C, and vice versa. For example, some genotypes may be resistant to certain infectious diseases, others may be able to get along on less food, still others give brave warriors, or peaceful citizens, or obedient slaves. A polymorphic, or genetically diversified, population which contains all these genotypes may have an adaptive advantage over a genotypically uniform one. The diversifying, also called "disruptive" (in my opinion, an unsuitable name), natural selection will maintain or increase the adaptive polymorphism. However, some or all of the genotypes from time to time may be confronted with environments unfavorable for them; such environmentally misplaced genotypes will act as constituents of the genetic load. This form of genetic load will be maintained in the population not by the pressure of recurrent mutation but by a balancing natural selection.

Another form of balancing natural selection can maintain a genetic diversity in a population even in a uniform environment. Suppose that the heterozygote, A_1A_2, for two gene alleles or gene complexes, A_1 and A_2, is heterotic, that is, adaptively superior to both homozygotes, A_1A_1 and A_2A_2. The heterotic form of natural selection will maintain both A_1 and A_2 in the population. This will be true even if one or both homozygotes are severely handicapped or lethal. The relative frequencies of A_1 and A_2 in a constant environment will depend upon the relative fitness of the two homozygotes; the speed with which these equilibrium frequencies will be attained will depend upon the degree of heterosis in the heterozygote. If the heterozygous and homozygous genotypes are clearly distinct phenotypically, we have the simplest form of balanced polymorphism.

This introductory article cannot describe several other forms of natural selection that maintain the genetic diversity in populations irrespective of mutation pressure. The major unsettled and controversial problem in population genetics is whether most genetic diversity in populations is maintained by the mutation pressure and counteracted by the normalizing natural selection, or whether an appreciable, and perhaps a major, part of this diversity is sustained by various forms of balancing natural selection.

The classical and the balance theories of population structure have rather different implications for studies on the genetic variables of human behavior. Suppose that the genetic diversity in human populations is principally mutational in origin, as contended by at least some partisans of the classical theory. Most individuals in a population would then be expected to be homozygous for "normal" alleles at most gene loci. The genetic diversity would be limited to a minority of loci, which in some individuals would be represented by at least one allele being a more or less deleterious mutant, waiting to be cast out by the normalizing natural selection. Although the mutant alleles might be numerous in the aggregate, the mutants at any one gene locus would usually be present in a small minority of individuals. It then would not be inconceivable that one would somehow apprehend the characteristics, including the behavior features, of that elusive entity, the carrier of the normal, typical, and optimal genotype of the species. All deviations from these typical features would then be accidental imperfections, to be eliminated as far as possible.

If the bulk of the genetic diversity in a species is balanced, the optimal genotype becomes a will-o'-the-wisp. At best, it is an abstraction convenient for some kinds of mathematical calculations. The genetic diversity will then be expected to be not only greater than the classical theory would suggest, but its character will also be different. No one genotype being the "optimal" or "normal" one, there is only an array of genotypes that make their carriers able to live and to reproduce successfully in the environments inhabited by the species. This array of genotypes may be designated as the adaptive norm of the species or population. It must be stressed that the "adaptive norm" is an array of forms not identical in their fitness in any one environment; some may be fitter than others in a certain environment, but less fit in other environments.

The ill-adapted, pathological variants are the expressed genetic load of the population, and the exceptionally vigorous or outstandingly successful or valuable variants, its genetic elite. Again, the role of the environment must not be lost sight of; a genotype belonging to the genetic load in one environment may shift into the adaptive norm in another environment. The same is true, of course, of the genetic elite. It is, then, the diversity of the genotypes and phenotypes, including the diversity of behaviors, and the factors that maintain the genetic variance in the populations, and not the "typical" nor even the statistically average behavior, that is most important to study.

It would be out of place to attempt here an evaluation of the evidence at present available for and against the classical and the balance theories of population structure. As stated above, the issue is a controversial one. It is, however, safe to say that the classical and the balance theories are not alternatives, in the sense that one of them must be wholly right and the other wrong. The problem is to what extent each of the two theories describes a real situation. It is undeniable that some genetic variants are maintained in populations by balancing selection, particularly by its heterotic form; it is likewise undeniable that some mutant genes are deleterious in all existing environments, both when homozygous and when heterozygous. The occurrence of such mutants in populations must be due to a mutation pressure.

Furthermore, no single solution of the problem can be valid for all living species. If, for example, it were demonstrated that most of the genetic variability found in populations of *Drosophila* is balanced, it would not follow that the same is true in human populations, or vice versa. Not only different species but even ecologically different populations of the same species may well contain different proportions of mutational and balanced variability. Comparative genetics is a line of study logically as well justified as comparative anatomy, comparative physiology, and comparative psychology.

GENETIC DIVERSITY IN EVOLUTIONARY PERSPECTIVE

Much of the foregoing discussion has dealt with genetic diversity in general, rather than with the genetic diversity in man. Genetic differences cannot be divided rigorously into purely morphological, purely physiological, and exclusively behavioral. It is a platitude that any genetic difference is necessarily a physiological and hence biochemical difference, because genes can act in no way other than through chemical messengers. The converse is not necessarily true—not every physiological difference is reflected in morphological or behavioral traits. Many, though not quite all, genetic differences produce what are known as pleiotropic, or manifold, effects; they alter two, or several, traits that do not have any obvious developmental interdependence. Darwin's classic example of pleiotropism (not called, of course, by this name) was that white cats with blue eyes are deaf. Many hereditary diseases, in man and elsewhere, produce complex syndromes of symptoms, and these syndromes often include more or less dramatic changes of behavior. Others are reflected in behavioral traits only very indirectly. For example, the ideal of feminine beauty has varied in different places and at different times all the way from linear to obese. That body build is genetically conditioned is not subject to doubt, but what behavioral consequences a given sort of body build will have depend on culturally developed tastes.

A developmental interdependence of traits may not be easily apparent, but it may exist nevertheless. There must be a reason why white fur and blue eyes go together with the deafness in cats. On the other hand, is the gene, or genes, that modify both the color of human hair and the color of the iris of the eye pleiotropic? The old idea that some forms of behavior in man go together with certain facial features and other "stigmata" has given rise to numerous superstitions, but the possibility of such correlations based on pleiotropic effects of genes should not be dismissed out of hand. The main point here is the need to be on one's guard against the hoary fallacy that there must be one gene for every trait and one trait for every gene. If you are puzzled why natural selection has established as species characters, and even as characters of higher categories, some apparently useless morphological traits, remember that the useless traits may be parts of pleiotropic syndromes that also include less conspicuous but more vitally important components.

The phenomenon of pleiotropism must be kept in mind in connection with the controversial problem of the so-called constitutional types, somatotypes, and racial types. As a geneticist sees it, the kernel of this problem is this: Are human populations polymorphic for genes, or for linked groups of genes sometimes called supergenes, that determine complexes of traits inherited as units? Sheldon and his followers claim that these complexes consist not only of constellations of morphological but also conjoined psychic traits. One could then imagine that there exist three genes, or gene complexes, forming multiple allelic series, which determine respectively the so-called endomorphic, mesomorphic, and ectomorphic "components." This hypothesis was submitted to a test only once, by Osborne and DeGeorge (1959). They somatotyped 59 pairs of identical and 53 pairs of fraternal twins, and had their typing confirmed by Sheldon. The heritability of the "components" proved to be low and, in addition, not equal in females and in males. This negative result may only mean, of course, that the criteria for the "components" have not been chosen correctly. Even more dubious are the claims of a Polish school of anthropologists (for example, Czekanowski, 1962, and Wiercinski, 1962), that they can distinguish in Polish and other European, and also in African, populations, a limited number of racial types inherited as units. No attempt to test the validity of this claim by using the twin method has been made, as far as I know.

As stated above, the genetic diversity that underlies the behavioral variations is not in a class by itself; it is similar in principle to that underlying the physiological and the morphological variations. On the other hand, it should always be kept in mind that the genetic and the evolutionary processes that control behavior have, in the human species, certain important peculiarities that make them different from those in any other species. It has already been pointed out above that one of the diagnostic attributes of the human species is genetically conditioned educability. The hallmark of humanity is the ability to learn, to profit by experience, and to adjust one's behavior accordingly. This developmental plasticity of human behavior is also man's most basic adaptive feature; without it, human social and cultural evolution would have been impossible. An individual whose genotype deprives him of this plasticity is an obvious misfit and belongs to the genetic load of the population in which he has appeared. The plasticity of learned behavior in man contrasts with the relative fixity of preponderantly instinctual behavior in other animals, especially insects. This does not mean that human behavior is independent of genetics; it does mean that the genetic system in man is constructed in a way that is unique in the living world.

The developmental fixity or plasticity of a trait is genetically conditioned. In general, a fixity is a property characteristic of traits or qualities, the presence and the precise form of which is indispensable for survival and reproduction. The developmental processes that give rise to such traits are said to be homeostatically buffered, or canalized; in other words, they can take place in about the same way over the entire range of the environments that the species normally encounters in its habitats. The environmental variations within that normal range do not influence the phenotype of the organism to any appreciable extent. With very few exceptions, babies are born with two eyes, a four-chambered heart, a suckling instinct, physiological mechanisms maintaining constant body temperature, and so forth. The genetics of developmental canalization has been studied experimentally in suitable materials, chiefly *Drosophila* flies. We can form a fairly good idea about how it is achieved in the evolutionary process.

There are, however, characters in which fixity would be disadvantageous, and a plasticity is beneficial. If a *Drosophila* larva is given ample food, it develops into a fly of a certain size. If the food is scarce, although above a certain minimum, the starving larva does not die; it pupates and gives an adult insect of a diminutive size. The number of eggs deposited by a *Drosophila* female on abundant food and at favorable temperatures may easily be ten times greater than with scarce food and an unfavorable temperature. A fixity of the body size and of the number of eggs produced would evidently be disadvantageous; instead of stabilizing these traits at certain values, natural selection has destabilized them, made them contingent on the environment in which an individual finds itself. Experimental evidence on how such a destabilizing selection may work is limited, but a good paper was recently published by Prout (1962).

The genetic endowment of *Homo sapiens* guarantees the development in all individuals belonging to the adaptive norm of the species of such capacities as educability, symbolic thinking, and communication by symbolic language. It is perhaps justified to say that human evolution has been dominated by a stabilizing selection for these capacities, and for a consequent destabilization of the overt observable behavior. This is responsible for the illusion that man at birth is a *tabula rasa* as far as his prospective behavioral development is concerned. In reality, the educability goes hand in hand with a genetic diversity. Any society has diverse functions and callings. With the development of civilizations the diversity of vocations has not grown smaller; on the contrary, it has increased, and continues to increase, enormously. The division of labor in a primitive society is chiefly between sexes and among different age groups; in a civilized society it is between individuals, occupational groups, guilds, social classes, or castes.

The trend of cultural evolution is obviously not toward making everybody have identical occupations, but toward an increasingly differentiated occupational structure. What is the adaptively valid response to this trend? First of all, the human genetically secured educability makes most individuals trainable for most occupations. Almost everybody, if brought up and properly trained, could become a fairly competent farmer, a craftsman of some sort, a soldier, a sailor, a tradesman, a teacher, or a priest. It is, however, eminently probable that some individuals, because of their genetic endowments, are more easily trained to be soldiers and others to be teachers. It is even more probable that only some individuals have the genetic wherewithal for certain specialized professions, such as a musician or a singer or a poet, or for peak achievement in sports or wisdom or leadership.

Human educability is traditionally emphasized by those who espouse liberal political views, while the genetic differences are harped upon by conservatives. As pointed out above, this is sheer confusion. The main tenet of liberalism, it seems to me, is that every human being, every individual of the species *Homo sapiens* is entitled to equal opportunity to achieve the realization of his potentialities, at least insofar as they are compatible with the realization of the potentialities of other people. Now, an equality of opportunity is needed not because people are all alike but because they are different. If they were all alike in their capacities and tastes, it would not really matter who is embarking on what career; the careers and vocations could as well be distributed by drawing lots. The diversifying form of natural selection, however, made the human populations polymorphic for tastes and capacities. A class or caste society leads unavoidably to the environmental misplacement of carriers of different genotypes. The biological justification of equality of opportunity is that it minimizes the chance of loss to the society of valuable human resources, as well as the chance of personal misery resulting from environmental misplacement.

ACKNOWLEDGMENTS

This paper went through several rewritings and revisions. I have benefited by discussions of the whole paper or of its parts with Drs. Alec Bearn, Ernst Caspari, E. B. Ford, and A. E. Mirsky, and with the members of the Conference on Behavioral Consequences of Genetic Differences in Man at Burg Wartenstein. Patricia Hall has patiently typed and retyped the consecutive versions.

BIBLIOGRAPHY

BRONOWSKI, J.
  1956. *Science and human values.* New York: Julian Messner.
COON, C. S.
  1962. *The origin of races.* New York: Alfred A. Knopf.
CZEKANOWSKI, J.
  1962. The theoretical assumptions of Polish anthropology. *Current Anthrop.*, 3: 481–494.
DOBZHANSKY, TH.
  1962. Genetic entities in hominid evolution. In S. L. Washburn (Ed.), *Classification and human evolution.* Chicago: Aldine.
  1963. *Mankind evolving.* New Haven, Conn.: Yale University Press.
  1964. *Heredity and the nature of man.* New York: Harcourt Brace.
DUNHAM, B.
  1964. *Heroes and heretics.* New York: Alfred A. Knopf.
FORD, E. B.
  1940. Polymorphism and taxonomy. In J. Huxley (Ed.), *The new systematics.* Oxford: Clarendon Press.
  1964. *Ecological genetics.* London: Methuen–John Wiley.
HUTCHINSON, G. E.
  1959. A speculative consideration of certain possible forms of sexual selection in man. *Amer. Natur.*, 93:81–91.

KANAEV, I. I.

1963. Essays on the history of comparative anatomy before Darwin. (In Russian) Moscow-Leningrad: Academy of Sciences U.S.S.R.

LOVEJOY, A. O.

1960. *The great chain of being.* New York: Harper.

MAYR, E.

1963. *Animal species and evolution.* Cambridge: Belknap Press.

OSBORNE, R. H., and F. V. DeGEORGE

1959. *Genetic basis of morphological variation.* Cambridge, Mass.: Harvard University Press.

PROUT, T.

1962. The effects of stabilizing selection on the time of development in *Drosophila melanogaster. Genet. Res. Camb.,* 3:364–382.

WIERCINSKI, A.

1962. The racial analysis of human populations in relation to their ethnogenesis. *Current Anthrop.,* 3:2–46.

ZIMMERMANN, W.

1953. *Evolution, die Geschichte ihrer Probleme und Erkenntnisse.* Frieburg-München: K. Alber.

INTELLECTUAL FUNCTIONING AND
THE DIMENSIONS OF HUMAN VARIATION[1]

JERRY HIRSCH

O NE DIFFICULTY in bridging the gap between the behavioral sciences and genetics lies in the very different concepts they apply to their subject matter. Behavioral science has been almost exclusively concerned with phenotypes, whereas the genotype-phenotype distinction lies at the foundation of genetics. Since it is now becoming evident that in many instances the same phenomena can be, and are being, studied by both disciplines, it has become important that we work toward a common frame of reference.

I shall argue that failure to understand fully the genotype-phenotype distinction seriously limits the growth of substantive knowledge in the behavioral sciences. It fosters the perpetuation of an unfruitful, and fundamentally metaphysical, dualism—a modern counterpart of the mind-body confusion. It has prevented the behavioral sciences from fully appreciating the difference between empirically empty formalisms, such as the laws of mathematics or the rules of chess, and models or relations with substantive content, such as the Bohr atom, the periodic table, the Watson-Crick model, Mendel's picture of heredity, the biochemical nature of dark adaptation, or the genotypic bases of individual and population differences.

There is an easy transition to dualism, either implicit or explicit, when behavioral science theory and research are based on what elsewhere has been proved to be the counterfactual uniformity assumption (Hirsch, 1962, 1963). When all organisms within a species (and very often across species, see Skinner, 1957, p. 368) are assumed to be essentially the same with respect to those features of their behavior whose laws are "reducible to statements of relationships between" behavior and "the environment within which it occurs" (Estes *et al.*, 1954, p. 270), the stage has been set for an empty-organism behavioral science in which the individuals "intervening between stimulus and response are [treated as] equivalent 'black boxes,' which react in uniform ways to given stimuli" (Erlenmeyer-Kimling and Jarvik, 1963, p. 1477). It is the uniformity assumption as much as anything that leads to dualism because of the rationale it supplies for the belief in the possibility of laws of behavior that are effectively independent of the specific organism that is behaving, that is, stimulus response relations of universal

1. Some of the ideas developed here have now been discussed in other contexts (Hirsch, 1967a, 1967b).

generality (if all organisms are alike, what holds for one must hold for others; therefore, why bother about "accidental" differences).

On this basis elaborate theoretical systems have been, and are still being, developed in which no reference whatsoever is made to the nature of the organisms whose behavior they purport to describe. Since this approach does not deny the reality of the natural world and its physico-chemical substate (that is, is not solipsistic) but, nevertheless, proceeds as if the realm in which behavior occurs were somehow divorced from the world of the natural sciences, in effect, it is just as dualistic in outlook as are explicit mind-body theories.

A viewpoint that avoids both implicit and explicit dualism has been well stated by the physicist Erwin Schrödinger in the discussion that helped set Maurice Wilkins on the road leading to the Watson-Crick model:

It is well-nigh unthinkable that . . . laws and regularities . . . should happen to apply immediately to the behavior of systems which do not exhibit the structure on which those laws and regularities are based (1946, p. 3).

THE GENETIC FRAMEWORK

Next, in a model that might facilitate generalization, I shall represent each gene geometrically by a dimension in a multidimensional coordinate space, where the permissible values along each dimension are the allelic alternatives of the gene it represents. Each chromosome contains hundreds, possibly thousands, of genes. Hence the genome of a gamete can be represented by a point in an n-dimensional space, where the letter n represents the total number of genes for a species and is fairly large, say anywhere from 10,000 to 50,000 in man (Spuhler, 1948). Since the genotype results from the union of two independently formed gametic genomes, it can be represented by a point that is the intersection of two n-dimensional genome spaces. Therefore, genotypically an individual is a single point in a hyperspace that defines the sample space of possible genotypes for a species.

The members of a population can be represented by a distribution of points in the hyperspace of genotypes. Since the dimensions in this hyperspace are independently variable and their number is very large, the probability of any two zygotic genotypes having all the same coordinates and thus occupying the same point in the sample space is vanishingly small. Furthermore, the set of points representing the genotypes of one generation will be replaced by a different set of points for the following generation. Consideration of the nature of the genotype shows that, although a population consists of sets of substitutable and replenishable points, each point is unique and nonreplicable. Therefore, the population may be represented over the generations by a cloud of ever-shifting points.

It has become increasingly clear of late that an understanding of the structure of biparental populations with the manifold multivalued dimensions of allelic variation that comprise their gene pools has fundamental importance to behavioral science. Since the lawful relations that make up the content of any science are an expression of the structure of the events studied, the more intimate our knowledge of population structure and of its physical basis becomes, the clearer will be our ideas about what relations are possible in the behavioral sciences. We now see that, at the level of the genotype at least, there are thousands of dimensions of variation, very many of which should be mutually independent.

BEHAVIOR

Since our primary concern at this time is with the relations between heredity and behavior, I shall attempt to define "behavior" as explicitly as possible. A survey of textbook and dictionary definitions reveals that more often than not its meaning is taken for granted. Furthermore, many definitions that do appear are unsatisfactory, because they usually hinge upon response to stimulation. As Skinner has pointed out (1938), while response to stimulation certainly does occur, so does behavior occur in the absence of antecedent events that bear an easily demonstrable relation of stimulus for a given behavior.

In the physical sciences, ever since J. W. Gibbs, a system has been defined as that part of the universe chosen for consideration. In the behavioral sciences we can define behavior as that part of the activity of an organism that we choose for consideration.

The study of behavior employs the descriptive methods of the naturalist, classificatory taxonomic methods such as are used by the medical diagnostician and the biological systematist, and the analytic experimental methods of the laboratory scientist. In Ernst Mayr's words, "it is the basic task of the systematist to break up the almost unlimited and confusing diversity of individuals in nature into easily recognizable groups, to work out the significant characters of these units, and to find constant differences between similar ones" (1942, p. 9). Similarly, the student of behavior must select, from the continuous stream of activity that is behavior, those units that are suitable for study and that can be classified.

The importance of a "suitable" choice of descriptive features and taxonomic categories will be appreciated when it is realized that the units of behavior chosen and the criteria by which they are recognized will strongly influence the future course of research with respect to both the kinds of analyses performed and the kinds of interpretations made. It will therefore be convenient to consider the study of behavior as consisting of three distinguishable phases: description, taxonomy, and analysis.

DESCRIPTION

To be part of science, nonintrospective observations must be formulated in terms that will place them in the public domain. An adequate description of a behavior presents a statement of its differentiating properties and of its relations to various events, so that other observers can identify it.

Descriptions of behavior may refer to either the means by which it is executed or the goals that it attains—what Hinde (1959) has called "physical description" and "description by consequence." Ideally, physical descriptions would be made in terms of the intensity, duration, and pattern of the physical activities of the body. In practice, though, to avoid cumbersome detail, only the grosser aspects of behavior are usually described, for example, the startle reaction, the amount of time spent in crouching by a bird, the rate of intromission in sexual behavior, and so forth. Descriptions in terms of consequences, made by reference to the goals of behavior, do not always include an account of the particular means by which these goals are sought. This type of description groups together into broad categories (such as reproductive behavior, homing, maze-learning) all those performances that achieve the same end state.

The two types of description, however, are not always alternative. In some cases, description in terms of muscular contractions is undesirable because of the unwieldy complexity of the data that result. In others, description by consequence is equally undesirable:

Sometimes—as in the threat and courtship posture of birds, which involve both a relatively stereotyped motor pattern and an orientation with respect to the environment—both must be used (Hinde, 1959, p. 571).

TAXONOMY

An ideal behavioral taxonomy would provide a theoretical system under which classes of behavior could be arranged in some meaningful and consistent way. The classes in the taxonomy would be defined in terms of the behaviors they include. In such a system the subjects of classification are behaviors, and the subjects of taxonomy are classifications (King and Nichols, 1960; Simpson, 1961). As yet, no taxonomic system has been devised that succeeds in doing for behavior what the Linnaean system attempted for species.

ANALYSIS

Analysis breaks behavior down into components that can be related to antecedent and contemporary events as well as to the structure of the organism. The responses in any behavior can be represented as points in a four-dimensional space whose axes may be labeled response, time, stimulus, and individuals. The experimental analysis of behavior may thus be thought of as the study of interrelations among these four dimensions, because behavior shows, at one and the same time, (1) temporal variations or differences in response over time, (2) stimulus-response covariation or changes in response as a function of changes in stimulation, and (3) individual variation or response characteristics that depend upon the particular organism observed.

Terms like conditioning, learning, maturation, fatigue, adaptation, sensitization, and so forth, refer to those aspects of behavior in which, for a given individual observed under constant stimulus conditions, a response changes over time. Tropisms, taxes, kineses, preferences, and all of the psychophysical relations refer to those aspects of behavior in which, for a given individual observed over a given time interval, the response changes as a function of changes in stimulation. Mental tests and psychometric methods are techniques for the measurement and analysis of individual differences in response; for all behavior, under a given set of stimulus conditions, over a given time interval and in a specified population, the characteristics of the response will depend upon the particular individual observed. The three major methodologies employed in the experimental analysis of behavior are conditioning, psychophysics, and psychometrics (after Hirsch, 1962, 1964).

UNITS AND DIMENSIONS

In animal ethology, there is a face validity to according natural-unit status to such behaviors as nesting, courtship, and predation—activities that have obviously been molded by the prolonged interaction of the species genome and the forces of natural selection. Man too is an animal whose characteristics have evolved through natural selection. But, as we now study his behavior in civilization, there is no comparable face validity permitting us to apply the label "natural" to most of the units we observe.

In contemporary behavioral science, far more attention is paid to man's social roles than to his biological properties. In industrial psychology, for example, tests are devised to select individuals who will most skillfully perform those tasks for which they are needed by industry. It would be folly to argue that man has gone through much natural selection for most of these specialized skills. On the other hand, man clearly does employ the capacities that he has evolved in the exercise of these skills. The great challenge now before the behavioral sciences lies in the analysis of man's biological properties and the elucidation of their modus operandi in a sociotechnological context. I anticipate that it is along these lines that we shall see realized the objectives of this conference, namely, an understanding of the behavioral consequences of genetic differences in man.

The discussion and analysis of phenotypic variation would become far more consistent if we treated as dimensions the behavioral characteristics in which individuals differ. Much awkwardness and confusion could be avoided if, instead of trait-bearers and nontrait-bearers (Fuller and Thompson, 1960, pp. 49, 70, 112, and so on), reference were made to the different modes or degrees of expression of a trait. While the language presently employed might appear adequate for distinguishing between tasters and nontasters, discussions of many characteristics like handedness, color vision, and the psychopathologies would be greatly clarified if all trait descriptions were uniformly cast in a dimensional frame of reference.

DIVERSITY AND ITS ANALYSIS

Near the end of the first book devoted exclusively to behavior genetics, Fuller and Thompson observe that

Possibly the most significant contribution of behavior genetics is its documentation of the fact that two individuals of superficially similar phenotypes may be quite different genotypically and respond in completely different fashion when treated alike (1960, p. 318).

While Fuller and Thompson's observation may be self-evident to students of evolution like Dobzhansky and Mayr who fully appreciate the population concept, it remains incomprehensible to the majority of behavioral scientists still trapped by the typological mode of thought. There is great heuristic value in making as explicit as possible the implications of their important observation. It certainly suggests that behavior study might offer the most sensitive means available for analyzing human diversity.

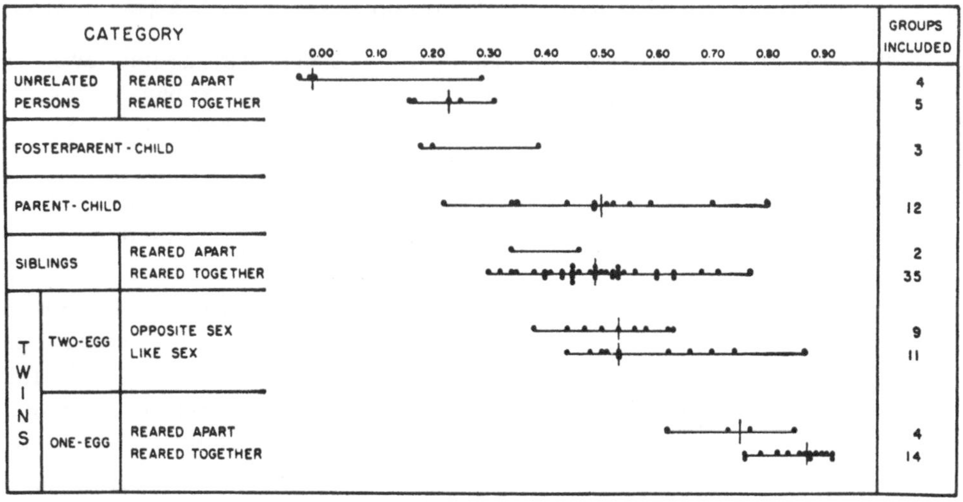

FIGURE 1

Distribution of "intelligence" correlations. Correlation coefficients for "intelligence" test scores from 52 studies. Some studies reported data from more than one relationship category; some included more than one sample per category, giving a total of 99 groups. Over two-thirds of the correlation coefficients were derived from I.Q.'s, the remainder from special tests (for example, Primary Mental Abilities). Midparent-child correlation was used when available, otherwise mother-child correlation. Correlation coefficients obtained in each study are indicated by dark circles; medians are shown by vertical lines intersecting the horizontal lines which represent the ranges. (Reprinted by permission from Erlenmeyer-Kimling and Jarvik, *Science*, 1963, 142:1477.)

A recent review of the literature on the relationship between heredity and tested intelligence revealed the remarkable fact that, for the distributions of correlational measures collected over the past half century, the central tendencies approach the values that would be predicted from a theory that intelligence is completely genetically determined (see Fig. 1). Like our understanding of evolution, meiosis, mitosis, and nervous conduction, this story too represents a synthesis of the work of many people:

The 52 studies . . . yield over 30,000 correlational pairings. . . . the investigators had different backgrounds and contrasting views regarding the importance of heredity. Not all of them used the same measures of intelligence, and they derived their data from samples which were unequal in size, age structure, ethnic composition, and socio-economic stratification; the data were collected in eight countries on four continents during a time span covering more than two generations of individuals. Against this pronounced heterogeneity, which should have clouded the picture, and is reflected by the wide range of correlations, a clearly definitive consistency emerges from the data (Erlenmeyer-Kimling and Jarvik, 1963, pp. 1477–1478).

This story supports, though certainly does not prove, the suggestion that behavior probably provides one of our most sensitive measures of human diversity. While I have indicated that the central tendencies of those distributions are consistent with the hypothesis of pure genetic determination, I do not nor do the authors of the review propose such an interpretation, any more than we should suggest that siblings are as alike genetically as parent and offspring just because their genetic correlations both happen to have the same average value.

The data suggest the sensitivity of performance measures to environmental conditions as well. There is a median correlation close to 0.23 for unrelated individuals reared together, a median close to 0.20 for foster parent-child, and a wider range of values for the parent-child correlations than for those between siblings. (A theory of complete genetic determination should predict no correlation among unrelateds as well as a greater variance for the sibling correlations than for those between parent and offspring.)

Clearly, data like those just reviewed merely demonstrate the heritability of a trait. That tested intelligence or most other human characteristics will show some measurable heritability is knowledge that should no longer evoke surprise. Since heritabilities are population, situation, and generation specific, studies that merely estimate their magnitude contribute knowledge of little significance at this time. Furthermore, the same applies to studies estimating the number of genes that can "account for" this or that complex trait, for example, intelligence (Pickford, 1949) or open field and Y-maze behavior (Fuller and Thompson, 1960, p. 268).

The truly challenging task now before us lies in the identification and behavior-genetic analysis of the phenotypic dimensions of human variation. In my opinion, they will not be identified through the exclusive study of complex behaviors, because we should expect them to be relatively simple, numerous, and largely uncorrelated. Estimates place the number of human genes at well in excess of 10,000. Over several generations and across a world population of more than three billion individuals, by the nature of the genome, very many of these genes must be assorting independently. It is therefore very unlikely that we shall learn much about their role in behavior by omnibus tests of intelligence or personality. The trouble with broad spectrum tests is that they measure too much. While they may be useful instruments of classification to serve the practical needs of society, because of their omnibus nature and their focus on social categories, we should not expect them to yield good measures of biological differences, which, because of the mosaic nature of the genome, should prove to be relatively fine grained.

Intelligence testing classifies individuals on the basis of their test performance:

(1) idiots score below 20; (2) imbeciles, 20–49; (3) morons, 50–69; (4) border-line deficients, 70–79; (5) dullards, 80–89; (6) normals, 90–140; and (7) geniuses, over 140. Since there are practically an unlimited number of ways of obtaining any score, lumping together all individuals who fall in the same category on cultural tests undoubtedly obscures many biological differences. When heritabilities are calculated, heredity and environment are interpreted as "accounting for" the estimated proportions of the variation over the range of test scores.

Forty-two years ago H. J. Muller studied "mental traits and heredity" (1925) in a pair of female monozygotic twins reared apart. His comments on those observations are still relevant:

The responses of the twins to all these tests—except the intelligence tests—are so decisively different almost throughout, that this one case is enough to show that the scores obtained in such tests indicate little or nothing of the genetic basis of the psychic make-up . . . it is necessary to institute an intensive search for ways of identifying more truly genetic psychic characters . . . despite the diverse reactions to almost all the non-intellectual tests used . . . there are really many other mental characteristics in which the twins would agree closely could we but find appropriate means of measuring them. Thus . . . the twins both seem possessed of similar energy and even tension, in their daily activity, with a tendency to "overdo" to the point of breakdown; both have similar mental alertness and interest in the practical problems about them, but not in remote or more purely intellectual abstractions and puzzles; both are personally very agreeable (as indicated by their popularity); both displayed similar attitudes throughout in taking the tests, even to such detail as lack of squeamishness in blackening the fingers for the fingerprints, and in being pricked for the blood tests—but turning away before the needle struck. The tastes of both in books and people appear very similar. It would seem, then, that the operations of the human mind have many aspects not yet reached by psychological testing, and that some of these are more closely dependent upon the genetic composition than those now being studied (1925, pp. 532–533).

Thus, there is reason to question the validity of prevailing testing procedures for the measurement of biologically significant properties both when heritability is present and when it is absent.

Another factor that has worked against our discovering the biologically relevant dimensions of behavioral variation has been the widespread preoccupation in the behavioral sciences with *general* processes like learning, perception, and motivation. The standard procedure has been to measure the performance of groups of subjects on specified tasks, average the resulting data, and infer the nature of the process from properties of the average curve. The subjects studied are usually obtained through schools, hospitals, industry, military installations, and other such institutions. Rarely are observations made on the relatives of these subjects. So, two of the conditions essential for uncovering the culturally significant biological dimensions of human variation are rarely, if ever, satisfied in the great bulk of behavioral science research: the careful analysis of human differences and the tracing through families of whatever segregation such differences might show.

THE DIMENSIONS OF HUMAN VARIATION

Both theory and observation point to the need for a radically different approach to behavior study. Theoretically, it is implied by our modern picture both of the mosaic organization of the individual genome and of the heterogenic nature of human populations. The complex of characteristics that comprises the total phenotype of each unique member of a population is the developmental result of thousands of genes, most of which, due to crossing over, undoubtedly assort independently.

Empirically, individual differences have been measured in the phenotypic expression of many traits and, for some, observations have been made on the similarities and differences both within and between families. On seven variables related to autonomic nervous function (vaso-motor persistence, salivary output, heart period, standing palmar conductance, volar skin conductance, respiration period, and pulse pressure) measured in children from three relationship categories (identical twins, siblings, and unrelated individuals), Jost and Sontag (1944) found that score similarity increased with genetic similarity. Fivefold within-family threshold differences have been found (von Skramlik, 1943) with taste stimuli that showed identical thresholds in monozygotic twins (Rümler, 1943). With somewhat more complex measurements, the fine structure of auditory curves was found to exhibit high intrapair concordance among identical twins, intra-familial similarities, and significant differences among unrelated subjects (von Bekesy and Rosenblith, 1950).

Individual differences in memory span have been measured among adult Caucasian men (Wechsler, 1952) as well as among Chinese boys and girls (Cheng, 1935). When attention span (immediately after oral presentation of a series of digits, demonstration by written reproduction of the number of digits remembered and introspective report that this performance did not involve mental grouping of the digits) and memory span (again, oral presentation and written reproduction, but with mental grouping permitted) were studied in the same individuals, both measures yielded well-dispersed arrays having a rather low correlation and very little overlap. The importance of the mental grouping operation is shown by the almost uniformly higher scores for memory span than for attention span (Oberly, 1928). With practice, some intelligent subjects can improve their memory span by learning to group items. Practice does not, however, seem to affect their attention span, which apparently sets a limit to the number of items they can combine into a group (Martin and Fernberger, 1929).

It is not at all uncommon to find that only some individuals in a diagnostic category receive extreme scores on specific trait measures. In 1914, Binet and Simon noticed the incapacity of some mental defectives to discriminate points on the skin, and they incorporated a test of this in their early intelligence measures (O'Connor and Hermelin, 1963). Birch and Mathews (1951) found poor auditory discrimination for tones above 6 KC in many, but not all, mental defectives. O'Connor (1957) found the incidence of color blindness in imbecile males to be higher than that in males of the general population. Berkson, Hermelin, and O'Connor (1961) found that the blocking of the EEG alpha rhythm, following the presentation of a bright light, lasted longer in some mental defectives than in normals. Siegel, Roach, and Pomeroy (1964) found that, following ethanol loading, some alcoholics showed significantly different plasma amino acid patterns from normals.

Many interesting family correlations can also be obtained. The relative paucity of available data can more likely be attributed to lack of interest in family studies on the part of behavioral scientists rather than to their nonexistence. Lennox, E. Gibbs, and F. Gibbs (1940) found that, although only 10 per cent of normal subjects showed occasional EEG abnormalities, some 60 per cent of the relatives of known epileptics had abnormal rhythms. Lidz, Cornelison, Terry, and Fleck (1958) reported marked distortions in communicating among many of the non-hospitalized parents of schizophrenic patients. McConaghy (1959), using an objective test for irrelevant thinking, found that at least one of the nonhospitalized parents of each of ten schizophrenic in-patients showed significant thought disorders. Of the twenty "normal" parents, twelve, or 60 per cent, scored within the range of responding characteristic of their offspring; whereas of 65 normal controls, only six individuals, or 9 per cent, scored in the range characteristic of the schizophrenics.

The most radical change in behavior study now being recommended involves a shift of focus away from insubstantial abstractions like learning, perception, and motivation to concern with consanguinity relations among the subjects we observe and to the study of the simplest units of intellectual functioning among individuals of known ancestry.

FUTURE PROSPECTS

In closing I should like to point out certain trends that are now developing. As the social, ethnic, and economic barriers to education are removed throughout the world and as the quality of education approaches a more uniformly high level of effectiveness, heredity may be expected to make an ever larger contribution to individual differences in intellectual functioning and consequently to success in our increasingly complex civilization. Universally compulsory education, improved methods of ability assessment and career counseling, and prolongation of the years of schooling further into the reproductive period of life can only increase the degree of positive assortative mating in our population.

Some might fear that this trend can only serve further to stratify society into a rigid caste system and that this time the barriers will be more enduring, because they will now be built on a firmer foundation. On the other hand, it may be noted that at least two conditions should prevent this from happening: (1) A significant contribution is undoubtedly made to intellectual functioning by the unique organization of each individual's total genotype and by its idiosyncratic environmental encounters. Furthermore, mutation, recombination, and meiotic assortment, plus our inability to transmit more than a small part of individual experience as cultural heritage, guarantee new variation every generation to produce the filial regression observed by Galton and to contribute to the social mobility discussed by Burt (1961). (2) The ever-increasing complexity of the social, political, and technological differentiation of society creates many new niches (and abolishes some old ones) to be filled by each generation's freshly generated heterogeneity.

BIBLIOGRAPHY

BERKSON, G., BEATE HERMELIN, and N. O'CONNOR
1961. Physiological responses of normals and institutionalized mental defectives to repeated stimuli. *J. Ment. Def. Res.*, 5:30–39.

BINET, A., and T. SIMON
1914. *The mentally defective child.* (Trans. W. B. Drummond) New York: Longmans Green.

BIRCH, J. W., and J. MATHEWS
1951. The hearing of mental defectives: its measurement and characteristics. *Amer. J. Ment. Def.*, 55:384–393.

BURT, C.
1961. Intelligence and social mobility. *Brit. J. Statist. Psychol.*, 14:3–24.

CHENG, P. L.
1935. A preliminary study of range of perception. *J. Testing* (in Chinese), II (2), 31 pp., in *Psych. Abstracts*, 1936, no. 5217.

ERLENMEYER-KIMLING, L., and LISSY F. JARVIK
1963. Genetics and intelligence: a review. *Science*, 142:1477–1478.

ESTES, W. K., S. KOCH, K. MacCORQUODALE, P. E. MEEHL, C. G. MUELLER, W. N. SCHOENFELD, and W. S. VERPLANCK
1954. *Modern learning theory.* New York: Appleton-Century-Crofts.

FULLER, J. L., and W. R. THOMPSON
1960. *Behavior genetics.* New York: John Wiley.

HINDE, R. A.
1959. Some recent trends in ethology. In S. Koch (Ed.), *Psychology: a study of a science*, Vol. 2. New York: McGraw-Hill.

HIRSCH, J.
1962. Individual differences in behavior and their genetic basis. In E. Bliss (Ed.), *Roots of behavior.* New York: Paul B. Hoeber.

1963. Behavior genetics and individuality understood. *Science*, 142:1436–1442.

1964. Breeding analysis of natural units in behavior genetics. *Amer. Zoologist*, 4:139–145.

1967a. Behavior-genetic, or "experimental," analysis: the challenge of science versus the lure of technology. *Amer. Psychol.*, in press.

HIRSCH, J. (Ed.)
1967b. *Behavior-genetic analysis.* New York: McGraw-Hill.

JOST, H., and L. W. SONTAG
1944. The genetic factor in autonomic nervous-system function. *Psychosomatic Med.,* 6:308–310. Reprinted in C. Kluckhohn and H. A. Murray (Eds.), *Personality in nature, society, and culture* (2nd ed.). New York: Alfred A. Knopf, 1953, 73–79.

KING, J. A., and J. W. NICHOLS
1960. Problems of classification. In R. H. Waters, D. A. Rethlingshafter, and W. E. Caldwell (Eds.), *Principles of comparative psychology.* New York: McGraw-Hill.

LENNOX, W. G., E. L. GIBBS, and F. A. GIBBS
1940. Inheritance of cerebral dysrhythmia and epilepsy. *Arch. Neurol. Psychiat.,* 44:1155.

LIDZ, T., A. CORNELISON, D. TERRY, and S. FLECK
1958. Intrafamilial environment of the schizophrenic patient: VI. the transmission of irrationality. *Arch. Neurol. Psychiat.,* 79:305.

MARTIN, P. R., and S. W. FERNBERGER
1929. Improvement in memory span. *Amer. J. Psychol.,* 41:91–94.

MAYR, E.
1942. *Systematics and the origin of species from the viewpoint of a zoologist.* New York: Columbia University Press. Reprinted by Dover Publications, New York, 1964.

McCONAGHY, N.
1959. The use of an object sorting test in elucidating the hereditary factor in schizophrenia. *J. Neurol. Neurosurg. Psychiat.,* 22:243.

MULLER, H. J.
1925. Mental traits and heredity. *J. Hered.,* 16:433–448.

OBERLY, H. S.
1928. A comparison of the spans of "attention" and memory. *Amer. J. Psychol.,* 40:295–302.

O'CONNOR, N.
1957. Imbecility and color blindness. *Amer. J. Ment. Def.,* 62:83–87.

O'CONNOR, N., and B. HERMELIN
1963. *Speech and thought in severe subnormality* (an experimental study). London: Pergamon Press.

PICKFORD, R. W.
1949. The genetics of intelligence. *J. Psychol.,* 28:129–145.

RÜMLER, P.
1943. *Die Leislungen des Geschmacksinnes bei Zwillingen,* Inaug. Diss., Jena. Cited in H. Piéron, 1962, *La Psychologie Différentielle* (2d ed.), p. 97, Paris: Presses Universitaires de France.

SCHRÖDINGER, E.
1946. *What is life?: the physical aspect of the living cell.* New York: Macmillan.

SIEGEL, F. L., MARY K. ROACH, and L. R. POMEROY
1964. Plasma amino acid patterns in alcoholism: the effects of ethanol loading. *Proc. Natl. Acad. Sci. U.S.,* 51:605–611.

SIMPSON, G. G.
1961. *Principles of animal taxonomy.* New York: Columbia University Press.

SKINNER, B. F.

1938. *The behavior of organisms: an experimental analysis.* New York: Appleton-Century.

1957. The experimental analysis of behavior. *Amer. Scientist,* 45:343–371.

SPUHLER, J. N.

1948. On the number of genes in man. *Science,* 108:279–380.

VON SKRAMLIK, E.

1943. Verebungsforschungen auf dem Gebiete des Geschmacksinnes, *Jenaische Zeitschr. für Medizin und Naturwissenschaft,* 50. Cited in H. Piéron, 1962, *La Psychologie Différentielle* (2d ed.), p. 97, Paris: Presses Universitaires de France.

VON BEKESY, G., and A. ROSENBLITH

1950. The mechanical properties of the ear. In S. S. Stevens (Ed.), *Handbook of experimental psychology.* New York: John Wiley.

WECHSLER, D.

1952. *The range of human capacities* (Rev. Ed.) Baltimore: Williams and Wilkins.

PSYCHOLOGICAL RESEARCH AND
BEHAVIORAL PHENOTYPES[1]

GERALD E. McCLEARN

THE RICH VARIETY of human behavior offers an essentially unlimited array of phenotypes for genetic investigation, and many diverse behavioral characters have already been investigated (see, for example, Fuller and Thompson, 1960). It is obvious, nevertheless, that only the most modest beginning has been made. There has recently been a striking increase in interest in behavioral genetics, and, in view of the accelerated research effort to be expected in the coming years, it appears appropriate to examine at this time some tactical and strategic considerations regarding the selection of phenotypes for future studies.

It is true, of course, that in an interdisciplinary area, such as behavioral genetics, choice of phenotype is sometimes determined by overriding considerations of theoretical or practical interest, wherein any shortcomings of the trait in terms of convenience of analysis are cheerfully tolerated because of intrinsic value. These considerations will naturally take priority. On the other hand, a range of phenotypes may be available within the general area chosen for investigation, and among these there may be large differences in efficiency and appropriateness from the point of view of genetic analysis. It is for this situation that the following comments are offered.

To begin, one might examine the more "successful" of the current phenotypes of human genetics to see if any general characteristic might be apparent. The first impression is that chance and serendipity play a large role in the discovery of phenotypes that are successful in the sense of opening new avenues of research or contributing fresh insights into the hereditary process. The discovery of PTC tasting, as is well known, was a fortuitous observation in a chemical laboratory (Fox, 1932). The description of phenylketonuria resulted from the recommendation of a mutual friend that the father of two defective children, who had described their unusual odor, consult Dr. Fölling, who was known to be interested in such things (Centerwall and Centerwall, 1963).

Even after discovery, it is not always immediately apparent that a phenotype is a promising one. The PTC taster-nontaster character was something of a curio for a while, then its usefulness in anthropological studies on race became apparent (see Allison and Blumberg, 1959), and a relationship to thyroid metabolism has now been investigated (see Fischer and Griffin, 1960). Landsteiner's (1901) brief

1. I am indebted to Professor Curt Stern and Professor William Meredith for helpful criticisms and suggestions.

paper on agglutination of normal human blood illuminated the therapeutic problems of blood transfusion, and suggested a possible forensic use in identification. But it could hardly anticipate the whole edifice of blood group genetics that has subsequently developed.

The delay in exploitation of a phenotype often is a matter of waiting for suitable innovations in technique. The explosive developments in cytological studies of Down's, Klinefelter's, Turner's, and other syndromes followed quickly the technique improvements of Hsu (1952) and of Tjio and Levan (1956), and the rather startling disclosure of the latter authors that the normal 2n complement in man is 46, rather than the long-accepted number of 48, chromosomes. Similarly, the development of biochemical techniques permitted Pauling and his coworkers (1949) to demonstrate the molecular basis of sickle-cell anemia, which had first been described almost four decades earlier. The whole area of study of hemoglobinopathies was thus opened up, and a consequence has been an enormous increase in detailed knowledge of gene action. As information on the genetics and biochemical nature of sickle-cell anemia and sickle-cell trait accumulated, Allison (1954) contributed a new direction to the research. His studies on malaria resistance of heterozygotes have made important contributions to the general problem of polymorphism in populations.

DESIRABLE ATTRIBUTES OF PHENOTYPES

To this point, examination of the "successful" phenotypes would suggest that the strategy to be recommended to a researcher is first to be lucky and thereafter to be patient. There are other generalizations to be drawn, however. It is noteworthy, first, that the "good" phenotypes have typically been minimally subject to environmental modification, and, second, that they have been qualitative categories, rather than quantitative scores. These points, it is to be feared, offer only limited direct guidance to the researcher in behavioral genetics, whose chosen traits most usually appear in a polygenically determined continuous distribution and are subject to alteration by environmental agencies. Nevertheless, together with certain other considerations, these characteristics of "good" phenotypes suggest at least a tentative set of criteria that may be applied to future work:

Variability. Although advances in molecular research may eventually make possible the chemical description of hereditary material that is invariant within a species, most genetic methods are suitable only for analysis of situations in which allelic differences exist. Thus the existence of individual variability at the phenotypic level is a prerequisite for research.

Adequacy of Measurement. The measurement involved in Mendelian characters is the nominal, or category-assigning, type of measurement. In cases where qualitative classes or categories are not recognizable and the trait appears to be continuously distributed, it is necessary to assign numbers reflecting varying degrees of the trait, and the assigned numbers may conform to an ordinal, interval, or ratio scale. In behavioral measurement, ordinal scales, in which only relationships of "greater than" or "less than" are indicated, are frequently encountered; ratio scales, implying the existence of an absolute zero of the trait, are practically never found. Interval scales, in which equal numerical differences in assigned score reflect equal differences in the trait being measured, are the aim of most behavioral researchers. They are difficult to achieve, however, and relatively few truly adequate interval scales are available in behavioral research.

The adequacy of the measurement scale has implications for interpretation of magnitudes or quantities of the traits being assessed and of statistical manipulations that can be employed (see, for example, Hays, 1963), but in all cases a prime requirement is that a high degree of reliability of measurement be obtained. While several types of reliability are distinguished by specialists, the type that corresponds to the genetical concept of repeatability, or stability over time, is the most relevant to the present discussion. It is necessary that the number assigned to an individual on one measurement occasion be the same, or nearly the same, as the number assigned on another measurement occasion.

From the point of view of measurement, then, a phenotype should be reliably measurable, and the closer the approximation to an interval scale, the better. A practical consideration might also be added in that ease and brevity of the measurement procedure is desirable.

Minimal Gene-Environment Correlation and Interaction. While the ease of genetic study is obviously greatest when environmental effects are negligible, to insist on this condition would be to eliminate study on many, perhaps most, theoretically interesting and socially important traits. The situation next best to one of no environmental effect is one in which the effects of given environmental conditions are equivalent for all genotypes, and environmental differences are distributed randomly in the population. Both of these conditions are difficult to assess. At least in certain traits, evidence is available to show that the conditions are not met. Drew, Colquhoun, and Long (1958), for example, have shown that the effects of one particular environmental agent, alcohol, on a simulated driving task are different for various personality (and possibly genetic) types. The correlation of genotype with environment seems large in traits such as intelligence where the parents most concerned with provision of good educational facilities at school and supportive measures in the home probably are those who, on the average, have also transmitted more alleles for superior intelligence.

Such failures to meet the assumptions of environmental additivity increase strikingly the difficulties of analysis and interpretation of genetic data. When possible, these difficulties should therefore be avoided.

Empirical and Theoretical Context. The value of a phenotype frequently emerges as its relationship to other traits and its physiological and biochemical correlates are elucidated. This suggests that a desirable feature of a to-be-selected phenotype would be a close relationship to an active field of empirical inquiry and theoretical formulation. Thus, many ancillary facts will already be known, and the process of accumulating information about related variables can proceed at a faster pace. Another benefit would be that developments in procedure and techniques, which, as we have seen, play an important role in exploitation of a phenotype, will be more likely to occur in areas that occupy the attention of other researchers, who may not necessarily be interested in the genetic aspects of the situation.

Appropriate Specificity of Trait. The advantages of the Mendelian character have already been mentioned. Behavior in the normal range is generally so complexly determined, with receptor, associative, and effector systems involved in even simpler behaviors, that a hypothesis of polygenic determination is, *a priori,* most probably correct in any given behavioral trait. This is not to say that monogenic conditions might not be found to affect the system under investigation. The homozygous recessive condition for phenylketonuria, which overrides the normal polygenic system determining intelligence, is a case in point. Much will unquestionably be learned from studying abnormal conditions such as this. But normal systems cannot be completely understood solely on the basis of abnormalities. Ultimately, therefore, the study of, say, intelligence is going to require dealing with a polygenic quantitative trait. But this does *not* mean that the most global measures possible must be taken. As will be discussed in somewhat more detail later, even such domains as intelligence and personality are divisible. It would appear reasonable that advantage in some degree would accrue from the use of the narrower (and presumably simpler) rather than the more extensive measures.

Having considered some of the desiderata of phenotypes, we may now turn to a brief examination of some appropriate areas of psychology in order to get some perspective on the wide variety of behavioral traits from which a choice can be made and on their relative strengths with respect to the criteria just discussed.

Since the time of Mendel, it has been the business of psychologists to measure behavior. They have become progressively better at it. A number of highly sophisticated techniques have been developed, and an enormous stockpile of information about a vast array of behaviors has accumulated. Much of this *terra* is *incognita* to geneticists in general, and even many human geneticists interested in behavior have not had the opportunity to examine more than a very small subplot of the whole. In the present paper, only a few areas can be discussed. Because I regard them as most appropriate, I have chosen the areas of testing, experimental, and differential psychology. Even with this restriction, nothing like an exhaustive or perhaps even a representative picture can be presented. It should also be noted that the boundaries between these areas are by no means hard and fast. At a given point in time, a psychologist might be using techniques or exploring problems relating to several or all of the areas mentioned. The distinction is primarily for didactic convenience. My intent is to characterize the areas briefly and to present some behavior traits from them that would appear to offer advantages for genetic research.

PSYCHOLOGICAL TESTING

In the latter part of the nineteenth century, several programs (notably those of Galton and Cattell) were aimed at measuring mental attributes. The tests employed included simple assessments of sensory and perceptual capacity such as pressure required to produce pain, hearing threshold, time estimation, and so on. In 1905, Binet and Simon combined some tests of this type with items requiring definitions of terms and reasoning. This test, which through many modifications and revisions became the well-known Stanford-Binet IQ test, was created in response to a practical social problem—that of identifying Parisian children who were feebleminded and thus not able to profit from instruction in the regular schools. Further pragmatic needs—those pertaining to mass assessment of the intellectual capacities of the soldiers conscripted in World War I—gave a further impetus to the field of test construction and resulted in the development of the Army Alpha and Army Beta tests. After the war, further tests were devised for the measurement of many aspects of intelligence, personality, aptitude, and interests.

Out of the practical needs a theoretical structure emerged, and highly sophisticated techniques were developed to cope with the problems of reliability and validity of measurement, establishment and interpretation of norms, multivariate analysis, sampling theory, control and standardization of conditions of testing, and so on. It is thus initially on the basis of adequacy of measurement that psychological test scores recommend themselves as phenotypes. In addition, of course, the development of a test presupposes variability within the population. The requirement of belonging within the context of a vigorous area is automatically satisfied by many of the tests, such as those on intellectual and personality functions, which have been central issues in psychology for many years. The advantages have been so obvious that tests have indeed provided human behavioral phenotypes from an early date. For example, Jones (1928) studied parent-child and sibling resemblance on Army Alpha and Stanford-Binet tests. Merriman's study (1924) used the Stanford-Binet, Army Beta, and National Intelligence Test, and subsequent researchers used some of the same or similar tests.

The principal theoretical issue during the development of the field of psychological testing has been whether intelligence could best be regarded as composed of one general factor with some specific factors contributing to special aptitudes, or as a composite of specific factors only. It would be inappropriate at this time (and for this writer) to discuss this matter in detail. For present purposes it is sufficient to note that the issue generated much research and resulted in the development of quite specific tests, measuring only a restricted aspect of intellectual functioning. Thurstone's well-known Primary Mental Abilities battery, for example, includes separate tests of *Verbal Comprehension, Word Fluency, Number* (computational speed and accuracy), *Space* (spatial or geometric perception, visualization of manipulations of objects in space), *Associative Memory, Perceptual Speed*, and *Induction* (general reasoning). Another well-known multiple aptitude battery is the Differential Aptitude Test, which yields scores on *Verbal Reasoning, Numerical Ability, Abstract Reasoning, Space Relations, Mechanical Reasoning, Clerical Speed and Accuracy*, and *Language Usage* (Spelling and Sentences). Other special aptitude tests have been developed for various sensory and perceptual processes, motor skills, mechanical, clerical, artistic, or musical aptitudes, literary appreciation, creativity, and so on.

In developments similar to those in intellectual function and aptitudes, though with somewhat less success, tests of personality have been constructed. One of the better known devices is the Minnesota Multiphasic Personality Inventory (MMPI) which provides scores for *Hypochondriasis, Depression, Hysteria, Psychopathic deviate, Masculinity-femininity, Paranoia, Psychasthenia, Schizophrenia*, and *Hypomania*. These scores have been determined so as to discriminate statistically between normal individuals and persons belonging to the diagnostic categories named. Another standard is the Guilford-Zimmerman Temperament survey, which gives scores on *General Activity, Restraint, Ascendance, Sociability, Emotional Stability, Objectivity, Friendliness, Thoughtfulness, Personal Relations*, and *Masculinity*.

These examples will serve to illustrate the wealth of specific measuring devices available. It was suggested earlier that much advantage might be obtained by more intensive investigation of these narrower traits rather than the more global ones. Support for this contention might be taken from the history of genetic studies on mental deficiency. As long as feeblemindedness was regarded as a single category, evidence was conflicting and controversy raged. Investigations of specific entities, such as Down's syndrome, phenylketonuria, amaurotic idiocy, and so on, have cleared the scene substantially, and, indeed, such conditions are now among the genetically best understood. While the tests of special aptitudes might not be able to measure a condition so unitary as these clinical entities, they may nevertheless provide a valuable improvement.

A more direct example may be taken from some recent human genetics research. Stafford (1961) presented evidence that on a test of spatial visualization (Identical Blocks Test), the various correlations of mothers or fathers with sons or daughters supported an hypothesis of sex linkage. Shaffer (1962) and Money (1963) have shown normal verbal IQ but depressed performance IQ values of patients with Turner's syndrome. Detailed analyses of subtest results and of factorial composition of the tests suggest that Turner's patients suffer from a degree of perceptual disorganization, particularly with reference to space-form perception. We thus find two separate lines of evidence implicating the sex chromosomes in spatial perception. Such findings, if confirmed, promise exciting new avenues of research, and they clearly could not have been made with global intelligence scores.

It would be unfair and inaccurate to convey the impression that previous research has completely neglected the analysis of components. Perhaps the best example to the contrary is the Hereditary Abilities Study discussed by Vandenberg in this volume and elsewhere (Vandenberg, 1962), which has presented a particularly detailed examination of behavioral traits, with 117 intelligence, aptitude, personality, sensory, perceptual, motor, and achievement tests.

It is undoubtedly correct, however, to suggest that we have only begun to realize the potential of the measurement devices available from the field of psychological testing.

EXPERIMENTAL PSYCHOLOGY

Experimental psychology is less properly defined by subject matter than by the general methodology of manipulation of independent variables, control or randomization of variables not under study, and measurement of dependent variables.

The variety of dependent variables measured is great, with responses ranging from muscle twitches to complex social behavior and with apparatus requirements ranging from paper and pencil to elaborate and complex electronic devices. Measurement must obviously be of at least minimal adequacy for the purposes of experimentation, and in many cases detailed studies of reliability and of scalar adequacy have been conducted on the measurement techniques.

The concern of a large proportion of studies in experimental psychology is to establish a significant mean difference between two (or more) groups. To this end, of course, within-group variance is an impediment. Perhaps for this reason curious attitudes have arisen toward individuality and its corollary of population variability. Variability is often treated as "error" that ideally should be eliminated by control procedures. Thus potentially valuable information about the distribution of traits in populations is often obscured by an exclusive concern with mean values, concern over variance being reflected solely in a measure of standard error. Nevertheless, scrutiny of the results reveals, as we expect, that the variability requisite for genetic study is indeed present.

It is with respect to amount of related material, however, that the behaviors studied in experimental psychology are particularly to be recommended. Many behavior domains have become centers of theoretical controversy, and the experimental information yielded in some cases is almost overwhelming. The area of verbal learning is a prime example. Research output in this area is so voluminous that a separate journal (now also overflowing) was initiated a few years ago. Replete with theoretical issues, dealing with material of the highest ultimate importance for education, and characterized by a wealth of highly sophisticated techniques and instrumentation, the field of verbal learning offers a cornucopia of phenotypes virtually untapped for genetic study. While verbal learning is admittedly an extraordinarily popular field of research and therefore not entirely representative, similar advantages may be expected from many other dependent variables of experimental psychology.

It is also possible that many such traits will be relatively free of problems of environmental confounding. For traits of personality and intelligence, the social values assigned lead almost inevitably to direct attempts to educate or discipline and thus to the serious complications of genetic analysis mentioned earlier. For many of the behavioral characteristics studied in experimental pyschology, no strong relationship to the socially valued traits is evident, and no direct environmental pressures are brought to bear. It is true that such traits may be related to others which are in fact subject to educational efforts and other environmental forces. However, some may be found to be comparatively free of features of environmental confusion and thus particularly valuable for certain types of inquiry.

To review representatively the field of experimental psychology is not possible here, but it is nevertheless desirable to convey some impression of the variety of phenotypes offered. The following examples are some that caught my attention as potentially valuable for genetic research in a brief survey. Most are well known and are described in standard reference works (see, for example, Osgood, 1953; Stevens, 1951; or Woodworth and Schlosberg, 1954).

Time Perception. Although less actively studied recently, time perception was the subject of intensive investigation in the early part of the century. Various aspects of accuracy of reproduction of temporal intervals, the degree of over- and underestimation, the effect of drugs, attitudes, and so on, have been investigated. A good review of the earlier literature is provided by Woodrow (1951). Large individual differences appear in a number of measures. For most persons in our culture, no obvious great advantage accompanies a high degree and no great disadvantage accompanies a low degree of skill in accuracy of temporal perception; no courses directly aimed at improving capacity in this respect are given in our educational system. Indeed, most people probably do not have the faintest notion of their proficiency in this regard. Whatever the environmental component of this trait, then, it would seem likely that genotype-environment correlations would be low.

Weight Discrimination. Discrimination between lifted weights has interested experimental psychology from its earliest days. It also would seem to be a trait with little genotype-environment correlation. In connection with this sensory-perceptual capacity, a whole series of psychophysical methods have been developed, and the adequacy of the measurement scales have been subjected to intense scrutiny.

Perceptual Constancies. The retinal image of a circular object such as a plate seen from an angle will be elliptical, yet the object is usually perceived as circular. The retinal image of a man at thirty yards distance will be smaller than that of an infant close at hand, yet the man is perceived as larger. These observations illustrate, respectively, the phenomena of shape and size constancy. These and other constancies relating to brightness, loudness, and color have been intensively studied. The adaptive value of constancies appears obvious: they permit the organism to respond to an object in terms of its objective characteristics over a broad range of circumstances.

Constancy is not an all-or-none phenomenon, and a frequently used measure is the "constancy ratio." If one were to respond solely on the basis of retinal size, for example, this would be represented by a constancy ratio score of zero. If the response is exactly appropriate to the objective size of the stimulus, the constancy ratio is 1.0. It is possible to obtain "overconstancy" ratios exceeding 1.0, representing overcorrections.

Individual differences in constancy are of substantial magnitude. Sheehan (1938) reported on constancy ratios for size, shape, and color constancies of 25 subjects. For size, the range was from about 0.5 to 1.0; for shape, from 0.5 to 1.25; and for color, from 0.4 to 0.95.

Other perceptual traits could be mentioned—perception of the vertical, auditory localization, figural aftereffects, closure, and so on, but the above will suffice for present purposes, and we may consider some other example from the area of learning studies.

Conditioning Phenomena. The extensive work in verbal learning has already been mentioned, but this represents only one facet of the research on human learning. Pavlovian conditioning techniques have been adapted to such responses as finger withdrawal from electrical shock and eyeblink to a puff of air. The various phenomena of extinction, generalization, differentiation, inhibition, and so on have been studied, and, naturally, individual differences appear in all of these measures.

Concept Formation. One type of situation employed in studying concept formation involves providing subjects with stimulus objects that differ from each other on several dimensions and instructing them to sort the stimuli into several categories. An interesting subject for investigation might be individual differences in tendency to categorize by number, size, shape, color, or some other dimension.

In this section, several expected advantages of phenotypes drawn from experimental psychology have been discussed, and a few specific examples have been cited. Typically the measurement techniques used have been appropriate to laboratory conditions. Most of them would be adaptable, however, for genetic investigations that might require modifications for such purposes as large-scale testing and testing in homes.

DIFFERENTIAL PSYCHOLOGY

The field of differential psychology is devoted to the study of the extent, nature, and causation of behavioral differences among individuals and groups. This field cuts across other classifications of psychology, and its behavioral bailiwick is thus as broad as all of psychology.

Individual Differences. The study of individual differences is usually categorized into the topics of intelligence, aptitudes, school achievement, personality, attitudes, interests, and perception; most of these have already been mentioned earlier. A particularly interesting extension of the study of individual differences is the relating of such differences to other behavioral responses. Utilization of individual differences in IQ, for example, has provided new perspectives on basic learning processes. Madsen (1963) showed that the widely announced and generally accepted "distribution of practice effect" appeared clearly for a low IQ group, but less clearly or not at all for average or high IQ subjects. Thus, a group of low (53–77) IQ subjects learned a task more readily when a one-minute rest was provided between learning trials than did a similar group given only a five-second intertrial interval. However, massed practice subjects did not differ significantly from distributed practice subjects in an average (93–112) IQ group or a high (129–166) group matched for chronological age with the low IQ group. Shapiro and Johnson (1965) matched subjects in mental age and again found that dull subjects, but not average or bright subjects, benefited by distribution of practice. Dent and Johnson (1964) showed the advantage of distributed practice for low IQ subjects of both "organic" and "familial" classifications. These results are particularly interesting because of parallel research which has shown that descendants of the selectively bred Tryon maze-dull rats were inferior to the maze-bright rats in learning a maze problem only under massed practice conditions, and were equal in performance when practice was distributed (Fehmi and McGaugh, 1961; McGaugh, Jennings, and Thompson, 1962).

Sex Differences. Differences between the sexes in a variety of aptitude and personality measures have been clearly demonstrated. While the distributions usually overlap extensively, males can be generally characterized as excelling in arithmetic reasoning, mechanical comprehension, and spatial orientation and spatial reasoning. Females excel in verbal fluency, dexterity, rote memory, numerical computation, and perceptual speed and accuracy (see Anastasi, 1958; Tyler, 1956). The interpretation of sex differences must involve sex-linkage, sex-limitation, differential cultural pressures, or a combination of such factors; the cultural explanation has long been the preferred one. It is easy, for example, to point out that sex differences in interests, attitudes, and personality conform to the sex roles of our culture. The causal relationship is not clarified by this fact, however. Whether the sex differences result from cultural roles or the cultural roles emerged from sexual differences has not been satisfactorily determined. Indeed, one would expect that the causal relationships are complexly intertwined. Closer scrutiny of these sex differences by genetic techniques are clearly warranted, and such investigations might open extremely valuable new research gateways.

Age Differences. In the study of age differences, differential psychology merges with the field of developmental psychology. Several features of this field have particular interest for the human behavioral geneticist.

Studies of intellectual functioning have shown that different subfunctions decline at different rates. Throughout middle age, vocabulary test scores show practically no decrement, motor skill loss is small, and visual perception and spatial relations ability show sharp declines. Detailed analyses of this nature should provide valuable leads about separate subentities of intellectual or personality functioning and thus provide suggestions about suitable phenotypes.

Another development has been the application of experimental techniques to the very young. The identification of stable patterns of individuality in neonates is most promising. In general, it is to be expected that problems of genotype-environment confounding will be reduced in measurements of infants on whom postnatal environmental factors will have had less time to operate.

A particularly productive researcher in this area is Hanuš Papoušek, whose paper in this volume should be consulted for details of his research results.

Class and Race Differences. In studies of differences among social classes and among races, differential psychology is dealing with some of the most urgent social problems of our time. The evidence is well known: overlapping distributions with significant mean differences are found between various socioeconomic groups, urban and rural populations, and racial groups. The difficulties of interpretation have also been carefully indicated: selective migration, educational handicap, difficulties of classification, culture conflict, the lack of truly culture-free tests, and so on. Detailed discussions of these problems occupy volumes, and any attempt to review the issues in this limited space would be hopelessly inadequate. Summaries of the differential psychological perspective on the problems of race may be found in Anastasi (1958) or Tyler (1956).

CONCLUSION

It has been the purpose of this paper to describe some areas of psychology, and to discuss the relative advantages from the standpoint of genetic research of the behavioral characters that constitute the subject matter of the areas. Perhaps we may draw the conclusion that something can be learned from the study of any phenotype, a substantial amount will be learned from systematic investigation of most phenotypes, and real breakthroughs will be derived from some few. The fields of psychology offer a large mine of phenotypes of varying complexity, ease of measurement, theoretical importance, and practical value. Within this mine are certain to be found some very rich veins.

BIBLIOGRAPHY

ALLISON, A. C.
  1954. Protection afforded by sickle-cell trait against subtertian malarial infection. *Brit. Med. J.*, 1:290–294.

ALLISON, A. C., and B. S. BLUMBERG
  1959. Ability to taste phenylthiocarbamide among Alaskan and other populations. *Hum. Biol.*, 31:352–359.

ANASTASI, ANNE
  1958. *Differential psychology.* (3d ed.) New York: Macmillan.

CENTERWALL, S. A., and W. R. CENTERWALL
  1963. The discovery of phenylketonuria. In F. L. Lyman (Ed.), *Phenylketonuria.* Springfield, Ill.: Charles C Thomas.

DENT, H. E., and R. C. JOHNSON
1964. The effects of massed vs. distributed practice on the learning of organic and familial defectives. *Amer. J. Ment. Def.*, 68:533–536.

DREW, G. C., W. P. COLQUHOUN, and H. A. LONG
1958. Effect of small doses of alcohol on a skill resembling driving. *Brit. Med. J.*, 2:993–999.

FEHMI, L., and J. L. McGAUGH
1961. Discrimination learning by descendants of Tryon maze bright and maze dull strains. *Psychol. Rep.*, 8:122.

FISCHER, R., and F. GRIFFIN
1960. Factors involved in the mechanism of "taste-blindness." *J. Hered.*, 51:182–183.

FOX, A. L.
1932. The relationship between chemical constitution and taste. *Proc. Nat. Acad. Sci.*, 18:115–120.

FULLER, J. L., and W. R. THOMPSON
1960. *Behavior genetics*. New York: John Wiley.

HAYS, W. L.
1963. *Statistics for psychologists*. New York: Holt, Rinehart, and Winston.

HSU, T. C.
1952. Mammalian chromosomes *in vitro*. The karyotype of man. *J. Hered.*, 43:167–172.

JONES, H. E.
1928. A first study of parent-child resemblance in intelligence. *Yearb. Nat. Soc. Stud. Educ.*, 27:61–72.

LANDSTEINER, K.
1901. Über Agglutinationserscheinungen Normalen Menschlichen Blutes. *Wiener klinische Wochenschrift*, 14:1132–1134. Translated in S. H. Boyer, IV (Ed.), *Papers on human genetics*, Englewood Cliffs, N.J.: Prentice-Hall, 1963.

McGAUGH, J. L., R. D. JENNINGS, and C. W. THOMPSON
1962. Effect of distribution of practice on the maze learning of descendants of the Tryon maze bright and maze dull strains. *Psychol. Rep.*, 9:147–150.

MADSEN, M. C.
1963. Distribution of practice and level of intelligence. *Psychol. Rep.*, 13:39–42.

MERRIMAN, C.
1924. The intellectual resemblance of twins. *Psychol. Monogr.*, 33: 1–58, Whole No. 152.

MONEY, J.
1963. Cytogenetic and psychosexual incongruities with a note on space-form blindness. *Amer. J. Psychiat.*, 119:820–827.

OSGOOD, C. E.
1953. *Method and theory in experimental psychology*. New York: Oxford University Press.

PAULING, L., H. A. ITANO, S. J. SINGER, and I. C. WELLS
1949. Sickle-cell anemia, a molecular disease. *Science*, 110:543–548.

SHAFFER, J. W.
1962. A specific cognitive defect observed in gonadal aplasia. *J. Clin. Psychol.*, 18:403–406.

SHAPIRO, G. M., and R. C. JOHNSON
1965. Personal communication.

SHEEHAN, MARY R.
1938. A study of individual consistency in phenomenal constancy. *Arch. Psychol.*, No. 222.

STAFFORD, R. E.
1961. Evidence for the sex-linked inheritance of spatial visualization. Paper presented at the American Society for Human Genetics, Atlantic City.

STEVENS, S. S. (Ed.)
1951. *Handbook of experimental psychology.* New York: John Wiley.

TJIO, J. H., and A. LEVAN
1956. The chromosome number of man. *Hereditas*, 42:1–6.

TYLER, LEONA E.
1956. *The psychology of human differences.* (2d ed.) New York: Appleton-Century-Crofts.

VANDENBERG, S. G.
1962. The hereditary abilities study: hereditary components in a psychological test battery. *Amer. J. Hum. Genet.*, 14:220–237.

WOODROW, H.
1951. Time perception. In S. S. Stevens (Ed.), *Handbook of experimental psychology.* New York: John Wiley.

WOODWORTH, R. S., and H. SCHLOSBERG
1954. *Experimental Psychology.* (rev. ed.) New York: Henry Holt.

BEHAVIOR GENETICS AND OVERPARTICULARIZATION: AN HISTORICAL PERSPECTIVE

DAVID A. RODGERS

B EHAVIOR GENETICS is a vigorously growing field in psychology that traces its ancestry with equal relevance through the historical antecedents of comparative psychology and individual difference studies. McClearn (1965), for example, documented the extent to which behavior genetics studies of animals have used the methods and focused on the problems of comparative psychology, while Hirsch (1962) detailed the central concern of this movement with individual differences. As Meissner (1965) described the modal emphasis, the focus is "on the determination of the genotypes corresponding to the psychological phenotypes. The overriding concern [is] to identify the genetic factors to which the observed variance in psychological abilities and performance could be ascribed" (p. 206). Similarly, McClearn (1960) described the task as one of determining "the relative contributions of genetic and environmental differences to the variability exhibited among individuals" (p. 149).

Some investigators have approached the task by selecting a class of behavior and studying the effects of genetic variations. For example, in early research using this approach, Sturtevant (1915) examined the effects of morphological mutations on sexual behavior of the *Drosophila*. Merrell (1949) and Williams and Reed (1944) carried forward this tradition in their *Drosophila* studies of effects of mutations on mating and on flight characteristics, respectively. Bruell (1962) studied the genetic variable of heterosis by means of preselected phenotypes, showing that several survival-relevant behaviors of hybrid mice are significantly enhanced over those of the parent inbred stocks.

Other investigators have preferred to start with known genetic constitution and examine the behavioral manifestations. Yerkes's pioneering studies of the dancing mouse (1907) were in this tradition, as was Thiessen's (1965) recent detailed study of the developmental characteristics of the wabbler-lethal mouse. Still others have related genetic variables to behavior by selective breeding, with phenotypic differences being the basis for systematic matings. Tolman's breeding of "bright" and "dull" rats (1924), later greatly extended and refined by Tryon (1940), provided the prototype for this approach. A variation of the selective breeding technique has been strain difference studies, in which the many available inbred strains of animals are systematically tested to identify stabilized genotypes that reliably differ according to some preselected behavior. The identification of mouse strains differing in preference for alcohol solutions (McClearn and Rodgers, 1959) is one of many examples.

Once systematic regularities are identified, the relative contributions of genetic and nongenetic factors to phenotypic variability are often examined, through studies of hybrid crosses. Mather (1949) detailed the techniques for one such approach, using genetically uniform strains as a starting point and analyzing relative variance in parent and first and second generation offspring. McClearn and Rodgers (1961) provided an example of this approach, in their analysis of the complexity of genetic factors underlying alcohol preference of mice. Bruell (1962) provided another example in his studies of several mouse phenotypes and mathematically elaborated Mather's approach as it specifically applies to behavior genetics research. Falconer (1960) detailed an alternative method, using heterogeneous populations, in which familial regressions take the place of comparisons of variance.

Enough work has already been completed within this tradition to make it abundantly obvious, as McClearn (1965) observed, that "genetic differences are fundamental to individuality, in behavior as well as in physical characteristics" (p. 91). This important principle, the initial *raison d'etre* of behavior genetics studies (Tolman, 1924), is now so well established that it no longer provides a satisfactory integrating influence or sense of direction for this rapidly growing discipline. The new goals being chosen appear to me to direct the movement toward a study of populations defined by their genetic characteristics in such a way that generalizations will be restricted to defined genotypes. The end point, as Hirsch (1962) observed, would be to "require experimental analysis for each behavior we wish to study for all populations in which we wish to study it" (p. 14).

This goal would counteract past tendencies, documented by Beach (1950, 1960) and Bitterman (1965), toward overgeneralization in a comparative psychology that has disregarded crucial subject differences (see Hirsch, 1962). At the same time, restriction of generalizations to defined genotypes would provide cogent methodological pressures for the field to focus increasingly on particularized units of behavior that have decreasing generality, eventually concentrating on "gene-specific" behaviors that uniquely reflect the action of single genes or that uniquely characterize specific genotypes (see Merrell, 1965). While sharpening and defining precise gene-specific units of behavior may have its descriptive, reductionistic, and even applied values, the increasingly narrow range of individuals to which generalizations could be made would markedly limit the theoretical and practical utility of such a trend. In a subsequent paper (Thiessen and Rodgers, 1967), the opinion is set forth that the goals of behavior genetics could desirably be redirected toward the study of "mechanism-specific" behaviors instead of behaviors uniquely associated with given genotypes. The present paper highlights the causes and problems of the overparticularization emphasis, through a selective review of historical antecedents.

THE COMPARATIVE BEHAVIORISTIC TRADITION

Robert Watson (1963) attributed to Aristotle the status of founder of psychology (p. 43). In two respects, partly because of subsequent historical developments, Aristotle has contributed profoundly to the present dilemmas of the behavior genetics movement. First, he systematically set man apart from other species as being uniquely capable of thinking and calculating behavior (Wheelwright, 1951, p. 152). Thus, a genetic classification, *Homo sapiens*, was made the exclusive basis for generalizations about a particular type of behavior. Direct continuity can be traced between this "species specific" emphasis and present tendencies to generalize according to genotype, tendencies that presage problems of overparticularization. Second, while portraying man as transcendent over all other species in possessing the "higher" mental processes, Aristotle also represented man as related to and sharing in the nature of all other forms of life. He explicitly stated, in Wheelwright's (1951) translation, that "certain traits in men differ only in degree from those in animals [and even where no identity or similarity is apparent] it is not unreasonable to regard animals as exhibiting traits analogous to those in man" (p. 113). Direct continuity can be traced between this emphasis and the contemporary behavioristic "black-box" fiction in comparative psychology that behavioral laws are (or should be) generalizable to all species regardless of individual differences. This fiction strongly predisposes toward problems of overgeneralization, the other horn of the comparative psychology dilemma from which the behavior geneticists in general and the overparticularists especially are fleeing. The present paper is a brief and highly selective historical account of these developments.

In broad conception, Aristotle hypothesized a pyramidal arrangement of souls (Boring, 1929, p. 156), with all living matter possessing vegetative or nutritive souls, all animal forms possessing sentient souls involving at least tactile qualities and usually many more sensory capacities, some animal forms possessing capacity for imagination, but only man possessing the ultimate pinnacle of psychic capacity, the ability to think, together with aspects of all the "lower" souls in the pyramid (a discussion of Aristotle's conception of souls is contained in several of his works, *De Anima* being perhaps the most comprehensive treatment).

Following the Dark Ages, when intellectual life was suppressed by superstition and when metaphysical conceptions largely supplanted mechanistic ones, Aristotle's writings became a primary basis for renaissance of rational thought acceptable to the medieval Catholic church. They therefore had great influence on cultural conceptions generally and have contributed substantially to the protopostulates of both academicians and laymen. Thomas Aquinas was perhaps the most influential scholar of this period of reinterpretation of Aristotle, during which a sharp dichotomy was drawn between soul and structure, or mind and body. This dichotomy was based on such comments of Aristotle as "the mind and . . . the thinking faculty . . . seems to be a distinct kind of soul and it alone admits of being separated, as the immortal from the perishable" (Hett, 1957, p. 77), and "mind is not intermittent in its activity. Its true nature becomes apparent, however, only when it is separated [from the lower functions] and is revealed as our only deathless and eternal part, without which nothing thinks" (Wheelwright, 1951, p. 148). Under the influence of Aquinas and other scholars of his time, Aristotle's supposedly uniquely human mind stuff, whereby man could think and reason and even contemplate himself contemplating, became explicitly associated with the Catholic church's metaphysical concept of the immortal soul and was sharply demarcated from the material body.

Descartes, in what might be called the beginning of the modern scientific era, further specified the duality of mind and body. He argued that they were separate although interacting entities, with the body being essentially a mechanical automaton without specific goal or purpose and the soul or mind being the quintessence of purpose which transcended all mechanistic considerations. This immaterial purposing psyche constituted the immortal soul that was the metaphysical concern of the church. By implication through omission, the dualistic position defined the mechanistic body as lying outside the interest of the church. Dualism thereby constituted a theological rationalization for scientific work on human physiology, providing an explicit philosophical basis for studying the body according to the same mechanistic conceptions that underlay other natural laws. It thus laid the groundwork for subsequent advances in physiology and physiological psychology.

Consistent with Aristotle's analysis, Descartes also emphasized the introspective and self-contemplative nature of the mind as its quintessential characteristic: "*cogito ergo sum*" (see R. I. Watson, 1963). He thus helped prepare the way for Wundt subsequently to anchor the experimental study of the mind on introspection.

In the immediate post-Cartesian period, the study of the mind was essentially philosophical and theological rather than experimental, while the scientific study of sensory physiology proceeded apace. Trading on the progress of physiology and changing theological attitudes, nineteenth-century scholars rescued mind as a legitimate area of scientific study in its own right. It was defined essentially by then as the content of consciousness, which was assumed to reflect the purposing, problem-solving, thinking behavior of the human animal. Although placing somewhat more emphasis on psychophysical parallelism than on Cartesian interactionism, these scholars retained the essential Cartesian dichotomy between mind and body. Such a dualism might not have been possible as the cornerstone of a new science, implying as it does the discontinuity between mechanistic and mentalistic principles, if an experimental procedure had not been devised which promised to allow systematic and objective study of conscious content. This procedure was, of course, trained introspectionism, which unabashedly took the contents of consciousness or "immediate experience" as observable data. The introspectionistic procedure in fact departed only slightly from the psychophysical studies of the sensory physiologists and was rationalized by long philosophical tradition that is well documented by Boring (1929), among others. Psychology was thereby established as the experimental science of the study of the mind, that uniquely human distillation of the purposive, problem-solving, controlling force of behavior.

Closely associated with this late nineteenth-century scientific approach to the mind and strongly reinforcing the Aristotelian interpretation of human mind as being at the pinnacle of the biological order was Darwin's (1859) brilliant genetic conception of "survival of the fittest." Logically, Darwin's position that all forms of life had evolved on the basis of natural selection from a single origin should have given rise to the conception of an inverted pyramid with the apex at the bottom and a broad base at the top, such that all current living forms were co-equal with all others in terms of demonstrated survival ability, each differing somewhat uniquely from all other forms. Instead, however, Darwin's work was interpreted to reinforce the Aristotelian conception of a pyramid with man at the apex. From this perspective, man was seen as having evolved beyond but through all other living forms, thereby retaining phylogenetic traces of "nutritive," "sentient," and "imaginative" soul stuff of all "lower" forms of life. All contemporaneous species were seen, from the myopic position of man's own achievements, as representing essentially preserved forms of the common ancestor beyond which man progressed at each bifurcation point in previous evolutionary history. Thus, all current living material in this conception merely represents stages in man's own evolutionary progress. This phyletic and Aristotelian view of man's biological contemporaries raised the intriguing possibility of finding antecedents of human characteristics in these so-called lower forms. The search for such antecedents became the preoccupation of comparative psychologists. As Beach (1950) observed, while decrying the condition, "the reactions of other animals have been of interest only insofar as they seem to throw light upon the psychology of our own species . . . With few exceptions American psychologists have no interest in animal behavior *per se*" (p. 119).

Darwin himself attempted to point out psychological continuities in terms of phylogeny (for example, 1872), but it was Romanes who was best known for his vigorous search for examples of human characteristics, particularly problem-solving behavior, in infrahuman animal forms. His "anecdotal method" based on observations and collected stories about pets and other animals (for example, 1882), which calls to mind contemporary accounts of "flying saucers," was effectively attacked by Morgan (for example, 1894), who argued that a given phenomenon might have many possible explanations and that, if there were alternatives, one should choose the simplest of these in preference to the more complex. This canon greatly influenced the early animal experimental work in psychology. Thorndike, among many others, implemented this conception in America by studying supposedly simple stimulus-response (S-R) models that offered promise of accounting for the learning of cats and other "lower" forms, and possibly for human learning as well.

The supposed parsimony of the S-R position is to be understood in part as a reaction against a mind-body dualism. Within the dualistic tradition, goal-oriented or teleological explanations of behavior, the primary alternative to the S-R position, implied an immaterial and unobservable mind stuff directing the actions, homunculus-fashion, of a body machine—Santayana's "clumsy conjunction of an automaton with a ghost" (Reid, 1962, p. 383). If the S-R position did not eliminate the homunculus (someone or something might manage the telephone switchboard between stimulus and response), it at least confined him within a "black-box" and tried to throw away the key. The subsequent development of teleological machines, obviously not incorporating the concept of dualism, has called into question the assumption that the S-R position is more parsimonious than one positing goal orientation regulated by negative feedback over the goal. The S-R approach nevertheless was a significant step toward laying the ghost of the immaterial mind. The expiating fee was that the organism intervening between stimulus and response, or at least the details of its information processing characteristics, was systematically ignored. The resulting successes of the S-R theorists, both in escaping the intellectual quicksands of the dualistic position and in systematizing behavior data, led to broad acceptance of the "black-box" fiction in comparative psychology, the assumption that behavior could be studied essentially independent of species or biologic considerations. Thus began the tradition of the psychological study of the laboratory rat, the tradition that has been documented by Beach (1950, 1960) and Bitterman (1965). It should be explicitly noted that the behaviors studied by these early psychologists, both before and after Morgan's canon and continuing after J. B. Watson's 1913 manifesto, were those behaviors that were assumed to reflect or systematically antedate human mental functioning. Thus, Wundt was studying human mind stuff directly and animal psychologists were studying its genetic antecedents and analogies, through the behavior they regulated, at lower levels of the evolutionary tree.

Calling both the introspectionists and the early animal psychologists to task in the same terms, J. B. Watson (1913) performed the phenomenal feat of redefining psychology to fit its methods rather than correcting the methodology to fit its avowed goals. Watson averred that both the trained introspectionists and the animal experimenters were observing stimulus input and behavioral output, not mind-stuff phenomena *per se*. He correctly argued that much purification and rigor would come from recognizing this, continuing the same experimental procedures, and openly accepting the study of "behavior" as the legitimate task of psychology while dropping the confusing and usually specious attempt to relate such behavior to internal mental process.

Over the next few decades, this behaviorist point of view was largely adopted. The various techniques of controlling stimulus and carefully observing response became the primary methodology of comparative psychology. Still, psychologists continued almost exclusively to study those forms of behavior that might reflect the operation of human mental processes. As a consequence, psychology, the so-called "science of behavior," largely neglected that vast array of behaviors that typify other species but lack any obvious analogy to human "mind-controlled" goal-oriented activity (see Beach, 1950, 1960; Bitterman, 1965).

Another move forward in conceptualizations in psychology was stimulated during the 1930's and since by a surprising assault on the behaviors that had specifically interested psychologists—that is, problem-solving behaviors—from an unexpected source tangential to zoology—ethology. This source demonstrated in a series of brilliant studies of releasor mechanisms that many complex behavior patterns that appeared to be "mind-guided" and problem-solving in nature would occur "automatically," without prior S-R association, if triggered by the appropriate environmental stimuli. For example, it was noted that the begging response of herring gull chicks depended on the patch of color on the mandible of the adult herring gull and could be elicited with a flat cardboard dummy having patches of similar color (Tinbergen, 1961). The frequency of response, in general, was a function of the similarity of this patch to that of the natural stimulus.

Many other examples of similarly "automatic," "unlearned" complex responses could be cited—for example, the sexual response of the three-spined stickleback (Tinbergen, 1951), the avoidance response of wild game chicks when exposed to a silhouette moved to resemble a hawk but not when it was moved to resemble a goose (Tinbergen, 1951), and the imprinting behavior of newly hatched ducklings (Hess, 1959). These studies appeared to reveal complex behavior sequences not easily distinguishable from patterns of human problem-solving activities and highly functional for species survival but that were far too rigid and too much under the control of "incidental" and "unlearned" releasor stimuli to be good analogs of human problem-solving. They did not fit explanations based on learned S-R associations and seriously challenged the "black-box" fiction that argues for the universality of behavior laws. These studies therefore strongly undercut the dominant rationale of comparative psychology.

Beach has been most explicit in trying to relate the implications of the ethologist movement to a redefinition of comparative psychology. He essentially rejected the "black-box" fiction and suggested a way in which comparative psychology could be freed from the hypocritical covert focusing on mind-stuff behavior while decrying the legitimacy of the mind as a scientific concept. Basically, he argued that behaviors are adaptive and the unique behaviors of each species are those that give it its differential advantages for survival. Such behavioral structures are laid down biologically and are of interest in their own right, not merely as analogs of human behavior. Thus, in two challenging papers a decade apart, Beach warned that the comparative psychology snark may be a boojum (1950) and called for the study of species-specific behavior (1960). This call for a more biologically based and more genetically defensible definition of the subject matter of comparative psychology was compatible with and gave stimulus to the growing interest of psychologists in behavior genetics.

THE INDIVIDUAL-DIFFERENCE TRADITION

Parallel to the comparative behavioristic tradition in psychology was the somewhat separate tradition of studies of individual differences. This tradition has also stimulated and found a congenial home in behavior genetics. If Darwin provided the key for the study of the phylogeny of behavior, which led systematically through the work of Romanes, Thorndike, and Watson to Beach's conception of species-specific behavior, then he also can be said to have stimulated the mental testing movement that originated with his cousin Galton and has constituted the central approach to the study of individual differences. The genetic emphasis in this movement was apparent from the beginning, as is reflected in the title to Galton's pioneering volume, *Hereditary Genius*, published in 1869.

While there has been a continuing interest in the genetic aspects of individual differences (for example in the eugenics movement), the mental testing movement in the United States was largely preempted by Binet's unitary concept of intelligence and its practical implementation through the useful general intelligence scales that Binet, Wechsler, and others developed. This unitary conception of intelligence fit neatly into the "black-box" and universality-of-behavioral-laws tradition of early twentieth-century psychology. It was almost successfully challenged by factor theorist L. L. Thurstone, with his demonstrations of "primary mental abilities." However, it was effectively defended philosophically and mathematically by Spearman, also a factor theorist, with his emphasis on a *g* factor. Such defense was no doubt made easier by the prevailing attitudes that favored general laws rather than particularized explanations.

Still another factor theorist, R. C. Tryon, working on the same problems of individual differences versus general intelligence, carried out the pioneering studies that ushered in the field of behavior genetics. In collaboration with Tolman, Tryon systematically bred strains of rats that showed markedly different capabilities to "learn" a maze (1940). This classical experimental demonstration of the possibility of gaining genetic control over precisely those phenotypes of central interest to comparative psychologists of the day sowed the seeds for what is rapidly becoming a revolution in psychology. Thus it was that behavior genetics was born out of the individual-difference movement.

THE EXTENT OF INDIVIDUAL DIFFERENCES
AS GENETICALLY DETERMINED

A brief consideration of genetic inheritance will indicate the incredible potential it has for insuring individuality. Each chromosome of genetic material presumably contains thousands of genes, each of which may exist in several allelic forms. Except in the case of closely related kinsmen, the probability becomes vanishingly small that any one person will possess two identical chromosomes or a chromosome that is identical to a corresponding chromosome of another person. Most marriages that do not involve recent common ancestors therefore bring together two people capable of transmitting 4^{23} possible combinations of the 23 pairs of chromosomes that constitute the human genetic material. In other words, a single married couple has the potential for producing approximately 70 trillion genetically unique offspring. Stated in another way, with the exception of monozygotic twins, the probability is about 1 in 70 trillion that any two children of a given marriage will be genetically identical (see Hirsch, 1962). The probability is vanishingly small that any two children of different marriages will be identical. This already astronomical potential for variability is further increased by a variety of additional genetic processes such as chromosome crossovers, mutations, and mosaicisms. Environmental factors, often in interaction with modifier and operator genes, constitute compelling additional pressures toward phenotypic variability, beginning with variations in intrauterine nutrition and continuing through the numerous other kinds of experiences that interact with the genetic material in the development of the individual (see Meissner, 1965; Rodgers, 1966).

In this almost infinite potential for variability lies a serious threat to both the species-specific conceptions and the behavior genetics movement. Intraspecies variability, fully as dramatic as interspecies variability, can be easily demonstrated. In one study, for example, over 97 per cent of variance in voluntary alcohol consumption was associated with strain differences within a single species, the house mouse (Rodgers and McClearn, 1962). One strain, the C57BL, showed consistent high alcohol intake, and other strains, such as the DBA, showed consistent avoidance of 10 per cent alcohol solutions. Why, then, should these strains not be taken as the basic units to be studied? But if intraspecies strains are to be preferred, then why not sublines within intraspecies breakdowns, until ultimately the problem is reduced to the study of the phenotypic expressions of single gene variations? Major phenotypic differences can be demonstrated between single-gene variations, even differences that uniquely influence "species-specific" responses (for example, Caspari, 1960); and much profit can come from careful attention to such material (for example, Thiessen, 1965). Nevertheless, there is a compelling reason why single-gene materials should not be the defining categories to be studied: it is a technical impossibility simply because there is an astronomical number of alternatives.

The precision gained through further purification and standardization of the genetic material is more than overweighted by the loss of ability to generalize to any useful range of subjects. However, such an ultimate reduction to highly precise impracticality is one logical extension of the standardized-genetic-material argument. There is evidence at hand that behavior genetics is oriented toward this *cul de sac*, suggesting the possibility that it may move inexorably toward an apparently unavoidable demise in a morass of infinite variability. For example, in commenting on the by now numerous demonstrations of strain differences in genetic manipulability of different behavior phenotypes, Caspari (1963) observed, "the answer does not lead us any further in our understanding of behavior. Furthermore, I cannot think of any meaningful question which would lead us further" (p. 98).

THE NEED FOR STRUCTURE-FUNCTION CLASSIFICATION

In this dilemma of overparticularization lies a curious preservation of the same flaw that has plagued psychology in the dualistic separation of mind and body. Cartesian mind-body dualism served a twofold function. On the one hand, it freed the body for experimental study; on the other hand, it set man's behavior apart, as had Aristotle centuries before, as something that was uniquely soul-guided and therefore different from that of other species. Thus, dualism classified behaving organisms (in the specific case, differentiated man from other species) by a system independent of structure-function relationships. The Darwinian conception of evolution reestablished continuity between man and other species, only to have subsequent events in the history of psychology overemphasize this continuity, in the form of the "black-box" fiction that ignores structural considerations. The species-specific and behavior genetics movements tend to correct the overgeneralizing that such systematic lack of attention to structure encourages, by proposing reproductive compatibility or genetic uniformity as defining classes within which generalizations are to be made. This genetic basis for classifying is still extrinsic to the structure-function relationships that are the ultimately irreducible units of behavioral analyses, however, just as is the anthropocentric classification that sets man apart by definition as possessing unique capacities. In effect, these positions all subtly argue that the basis for classifying behavioral regulating mechanisms (the "mind" in Cartesian terminology) should be extrinsic to those structural-functional relationships by which behavior is effected (the "body" in Cartesian terms). The flaw underlying mind-body dualism is thus preserved, even though the dualism as such is not retained.

It is my contention that the basic parsimonious unit of "explanation" in all sciences is ultimately a detailed description of structure, such that the functional relationships are an inevitable concomitant of the structure. This is indeed a point of view that Aristotle strongly advocates for all subhuman explanations but wavers from in considering the nature of the human psyche. It was his wavering, apparently based on theological considerations of his day and specifically utilized for subsequent theological concerns, that was crystalized into the influential mind-body dualism conception. The transition from nonstructural to structure-function emphases has marked the critical point of maturity for many sciences. For example, chemistry came of age when it passed from the magic of alchemy to the concept of atomic structure that underlies the periodicity of the elements. Genetics matured rapidly as the central integrating science of biology after it passed from the mathematics of Mendel to the mechanics of the chromosomes and the biochemistry and mechanics of the DNA molecule. Medicine became a major and respected contender in the scientific community as and to the extent that it gave up demonology and exorcism in favor of bacteriology and physiology.

A major advantage of structure-function analyses is the ultimate accuracy and parsimony with which principles can be generalized across a wide range of situations. This principle of generalizability of mechanism-specific, or structure-function, relationships has contributed greatly to the power of explanation in other sciences. For example, it is illustrated in the application of the law of the lever with equal relevance to the functioning of the malleus of the middle ear and the functioning of a crowbar. The structural-functional characteristics of these two devices—both are rigid bodies pivoted at one point and subject to activating and resistive forces at other points—define them as functionally equivalent with respect to the law of the lever, even though they differ grossly in other respects. In exactly parallel fashion, structural similarities, *when the relevant structures are identified and specifically related to function,* can become the defining bases for categorization of those organisms to which generalizations of findings about behavior can be made. In some cases, equivalent structure can reflect common genotypic influence. Indeed the capacity of the genome to produce predictable subsequent structure constitutes its appeal to the behavioral researcher. However, in addition to such genome-specific determination of structure, homologus, convergent, and parallel genetic determination of functionally equivalent structure across genomes and across species is also possible (King and Nichols, 1960). Genetic introduction of structural differences within and across species is also possible, of course. Nevertheless, environmental events such as imprinting experiences, nutritional factors, or even surgery or prostheses can both introduce and eliminate structure-function equivalents at variance with "genetic" expectations. Because of these commonly encountered variations, "mechanism-specific," or structure-function, analyses would seem to provide a more suitable basis for generalization than would either classification according to genetic uniformity or "classification" according to the "black-box" fiction that ignores structural characteristics altogether.

A subsequent paper (Thiessen and Rodgers, 1967) presents more explicitly the proposal that behavior genetics is a methodology uniquely suitable for studying structure-function, or mechanism-specific, relationships, thereby escaping the dilemma of extrinsic classification that leads almost inevitably to over- or undergeneralization. The present paper has attempted to analyze the development of the current emphasis in behavior genetics that, in Ginsburg's (1958) words, "brings us to a locked door."

BIBLIOGRAPHY

BEACH, F. A.
   1950. The snark was a boojum. *Amer. Psychol.,* 5:115–124.
   1960. Experimental investigations of species-specific behavior. *Amer. Psychol.,* 15:1–18.
BITTERMAN, M. E.
   1965. Phyletic differences in learning. *Amer. Psychol.,* 20:396–410.
BORING, E. G.
   1929. *A history of experimental psychology.* New York: Appleton-Century-Crofts.
BRUELL, J. H.
   1962. Dominance and segregation in the inheritance of quantitative behavior in mice. In E. L. Bliss (Ed.), *Roots of behavior.* New York: Hoeber-Harper.
CASPARI, E. W.
   1960. Genic control of development. *Perspectives Biol. Med.,* 4:26–39.

1963. Genes and the study of behavior. (Vice-presidential address.) *Amer. Zoologist,* 1963, *3,* 97–100.

DARWIN, C.
1859. *On the origin of species by means of natural selection, or the preservation of favoured races in the struggle for life.* London: J. Murray.
1872. *The expression of the emotions in man and animals.* London: J. Murray.

FALCONER, D. S.
1960. *Introduction to quantitative genetics.* New York: Ronald Press.

GALTON, F.
1869. *Hereditary genius.* New York: Macmillan.

GINSBURG, B. E.
1958. Genetics as a tool in the study of behavior. *Perspectives Biol. Med.,* 1:397–424.

HESS, E. H.
1959. Imprinting. *Science,* 130:133–141.

HETT, W. S.
1957. *Aristotle on the soul, Parva Naturalia, on breath.* Cambridge, Mass.: Harvard University Press.

HIRSCH, J.
1962. Individual differences in behavior and their genetic basis. In E. L. Bliss (Ed.), *Roots of behavior.* New York: Hoeber-Harper.

KING, J. H., and J. W. NICHOLS
1960. Problems of classification. In R. H. Walters, D. A. Rethingshafer, and W. E. Caldwell (Eds.), *Principles of comparative psychology.* New York: McGraw-Hill.

McCLEARN, G. E.
1960. Behavior and heredity. In *McGraw-Hill encyclopedia of science and technology.* New York: McGraw-Hill.
1965. The inheritance of behavior. In T. E. McGill (Ed.), *Readings in animal behavior.* New York: Holt, Rinehart, and Winston.

McCLEARN, G. E., and D. A. RODGERS
1959. Differences in alcohol preference among inbred strains of mice. *Quart. J. Stud. Alcohol,* 20:691–695.
1961. Genetic factors in alcohol preference of laboratory mice. *J. comp. physiol. Psychol.,* 54:116–119.

MATHER, K.
1949. *Biometrical genetics.* New York: Dover.

MEISSNER, W. W.
1965. Functional and adaptive aspects of cellular regulatory mechanisms. *Psychol. Bull.,* 64:206–216.

MERRELL, D. J.
1949. Selective mating in *Drosophila Melanogaster. Genetics,* 34:370–389.
1965. Methodology in behavior genetics. *J. Hered.,* 56:263–266.

MORGAN, C. L.
1894. *An introduction to comparative psychology.* London: Scott.

REID, J. R.
1962. The myth of Dr. Szasz. *J. nerv. ment. Dis.,* 135:381–386.

RODGERS, D. A.
1966. Factors underlying differences in alcohol preference among inbred strains of mice. *Psychosom. Med.,* 28:498–513.

Rodgers, D. A., and G. E. McClearn
1962. Mouse strain differences in preference for various concentrations of alcohol. *Quart. J. Stud. Alcohol*, 23:26–33.

Romanes, G. J.
1882. *Animal intelligence*. New York: Appleton.

Sturtevant, A. H.
1915. Experiments in sex recognition and the problem of sexual selection in *Drosophila. J. Anim. Behav.*, 5: 351–366.

Thiessen, D. D.
1965. The wabbler-lethal mouse: a study in development. *Anim. Behav.*, 13:87–100.

Thiessen, D. D., and D. A. Rodgers
1967. Behavior genetics as the study of mechanism-specific behavior. In J. N. Spuhler (Ed.), *Genetic diversity and human behavior*. Chicago: Aldine.

Tinbergen, N.
1951. *The study of instinct*. London: Oxford University Press.
1961. *The herring gull's world*. New York: Basic Books.

Tolman, E. C.
1924. The inheritance of maze-learning ability in rats. *J. comp. Psychol.*, 4:1–18.

Tryon, R. C.
1940. Genetic differences in maze-learning ability in rats. *Yearb. Nat. Soc. Stud. Educ.*, 39 (1):111–119.

Watson, J. B.
1913. Psychology as a behaviorist sees it. *Psychol. Rev.*, 20:158–177.

Watson, R. I.
1963. *The great psychologists from Aristotle to Freud*. Philadelphia: J. B. Lippincott.

Wheelwright, P. (Trans.)
1957. *Wheelwright's Aristotle, Natural science; the metaphysics; zoology; psychology; the Nicomachean ethics; on statecraft; the art of poetry*. New York: Odyssey Press.

Williams, C. M., and S. C. Reed
1944. Physiological effects of genes: the flight of *Drosophila* considered in relation to gene mutations. *Amer. Natur.*, 78:214–223.

Yerkes, R. M.
1907. *The dancing mouse*. New York: Macmillan.

BEHAVIOR GENETICS AS THE STUDY
OF MECHANISM-SPECIFIC BEHAVIOR

DELBERT D. THIESSEN AND DAVID A. RODGERS

BEHAVIOR GENETICS is rapidly gaining acceptance as a substantive discipline within psychology. In the 1965 *Annual Review of Psychology*, Bell observed that "the burgeoning and productive field of behavior genetics in animals hardly seems in need of advocacy" (p. 3). In the 1966 *Annual Review*, McClearn and Meredith called the growth "explosive." The direction of growth of this vigorous infant is already a matter of concern and controversy. For example, Caspari (1963) comments that the most direct approach to behavior by genetic means will "not turn out to be very promising." He is referring to the establishment of differences in behavioral characteristics between strains or sublines and subsequent analysis of their genetic behavior in crosses, a common paradigm of behavior genetics studies.

Commenting on the same trend, Ginsburg (1958) observed that such an approach "brings us to the threshold of a locked door." Notwithstanding these misgivings, the main emphasis of the behavior genetics movement still seems to be, as Meissner has recently expressed, "on the determination of the genotypes corresponding to the psychological phenotypes. The overriding concern [is] to identify the genetic factors to which the observed variance in psychological abilities and performance could be ascribed" (1965). These goals may be partly a reaction against past tendencies toward overgeneralizations that are based on genetically heterogeneous colonies of laboratory rats and that assume biological differences of behaving organisms are of little consequence. Commenting on such overgeneralization, Bitterman (1965) wrote, "It is difficult for the non-specialist to appreciate quite how restricted has been the range of animals studied in experiments on animal learning because the restriction is so marked."

The present focus would appear to overcorrect in the direction of emphasis on individual differences, however, with attention shifting from species-specific behavior (Beach, 1960) to strain-specific behavior to substrain-specific behavior and ultimately to gene-specific behavior (Merrell, 1965). Bitterman (1965), for example, argued for a phyletic table, with each phylum, class, or species (as the data require) taken as the independent dimension according to which evolutionary behavioral categories are ordered. Hirsch (1962) argued that each population of organisms constitutes a separate reference group calling for independent verification of behavioral laws. Ginsburg (1958), in turn, wrote that attention to specific gene action is the step that will unlock the door to further progress in behavior genetics. Meissner (1965) went further, in emphasizing the appropriateness of distinguishing between structural and regulator genes in assessing genetic correlates of behavior.

The main thrust is thus toward increasing fractionation of the reference groups to which generalizations can be made, with genetic uniformity or a combination of genetic and environmental uniformity being the basis for identifying reference group members. In the present paper, this trend is assessed as leading to fatal overparticularization. The alternative proposal is suggested that the techniques of behavior genetics can be utilized to isolate, identify, and clarify structural mechanisms that are specific to particular units of behavior wherever they appear. In brief, we argue for the use of behavior genetics to study *mechanism-specific*, rather than species-specific or gene-specific, behaviors.

THE EXTENT AND IMPORTANCE OF
INDIVIDUAL DIFFERENCES

Individuality is a characteristic of nature. The genetic, evolutionary, and experiential history of mammalian species, especially, underwrite phenotypic variability to an incredible extent. Even if concern is with only one genetic locus with two alternative alleles, say, A and a, three genetic classifications are possible. In general, for n alleles at a locus the population potential is $n(n+1)/2$ genotypic classes. When genomes are considered, much variation can be found among the diploid chromosomes. If two chromosome pairs are considered, four different genomes are possible. In general, for n chromosome pairs there exists a potential population of 2^n genomes. Thus, without reference to other genetic mechanisms, each man with his 23 pairs of chromosomes has the inherent variability of 2^{23} genomes. Since each parent shares this potential, the odds that any two of their offspring will share exactly the same genome (monozygotic twins excepted) is $(1/2^{23})^2$, or about one chance in 70 trillion (Hirsch, 1962). This remarkable range of variability is extended by chromosome crossover during the metaphase of meiosis, by allelic and non-allelic interactions, and by more complex mechanisms such as enforced heterozygosis, coadapted systems, linkage relations, position effects, and the like. To these potentials for variability must be added mutations, chromosome abnormalities, translocations, and mosaicisms. Chances for phenotypic variation must be some function of these genetic differences.

Of paramount importance in directing phenotypic expression is the developmental plasticity gained through physiologic interaction at all levels from the gene onward. Such plasticity is an interacting complex of gene-specific chemical reactions and intermediate reactions, intertwined with stimulus events (Waddington, 1957; Meissner, 1965). External events of consequence that are known to canalize gene penetrance or expression encompass early experiences of a diverse nature (De Fries, 1964; De Fries, Weir, and Hegmann, 1965), momentary stimulus fluctuations (McClearn, 1960) and social reactions (Thiessen, Zolman, and Rodgers, 1962). In all, it can be judged that the number of phenotypic outcomes is a function of the number of genotypes involved, the number of environments specified and the number of expressions that are possible in each environment.

In addition to the genetic diversity of individuals and the gene-environment relations during ontogeny, there is a diversity of another sort—that of gene frequency in the population. Mechanisms operative to maintain genetic and phenotypic differences within and between populations include everything from mutations and random drift to assortative mating and differential selection (Falconer, 1960). If this were not enough, behavioral variability itself may be a property of specific genetic units, so that response diversity in fluctuating environments can compensate for less mutable diversity at the level of the gene.

Views of the ubiquity of individual differences are prophetic. Those who study only common properties of behavior, or general behavioral laws, are severely pressed for an explanation of variations. Those who study species-specific behaviors cannot adequately handle individual differences within a species. And those who confront these differences are seemingly forced into a *cul-de-sac* of recognizing a unique controlling mechanism for each separate unit of behavior. In Hirsch's (1962) opinion the establishment of "laws" of behavior "will require experimental analysis for each behavior we wish to study for all populations in which we wish to study it" (p. 14).

This is not just a program outline, it is a recognition of and a concession to the uniqueness of any behavioral measurement. Where, then, are the common points of interaction between those who are determined to tease out of this welter of variability nomothetic characteristic of behavior and those who dissect generalities into idiographic particulars? The discord is recognized by behavior geneticists. As Hirsch (1958) said earlier:

Experimental psychologists have been so deeply involved with the invariant aspects of behavior processes that they have ignored ID's [individual differences] almost as an item of faith. Differential psychologists, on the other hand, have been so closely identified with variant test populations that they have scarcely had time for behavior processes (p. 6).

DEMONSTRATIONS OF GENETIC INFLUENCE

We will illustrate the usual methods of behavior genetics in dealing with individual differences and attempt to discern a trend in this area of investigation that might offer some hope for specifying general laws of behavior.

The remoteness of the behavior from the primary gene-polypeptide action has led to the assumption that there is a great deal of noncongruence between genetic and behavioral organization (see Scott and Fuller, 1963). This might have been a convincing argument for the exclusion of genetics from the study of behavior had it not been for the efforts of a small number of investigators who continually pointed out gene-behavior relations. In this regard, perhaps the greatest contribution of behavior genetics to the science of behavior has been the demonstration that nearly all behaviors studied are heavily influenced by the individual's genotype (see Fuller and Thompson, 1960).

Most striking have been the many observations of large differences in behavior among strains of mice known to differ genetically (Jay, 1963) and presumed to be nurtured in similar environments. Observations range from studies of general activity and emotion (Lindzey, Winston, and Manosevitz, 1963; Thiessen, 1965a) to those of perception and learning (Collins, 1964; McClearn, 1960). In fact, the pervasiveness of such differences is becoming the *sine qua non* of adequate scaling techniques in the measurement of behavior. Pedigree analyses in a similar fashion have suggested the influence of gene action. Such influences are particularly evident when single gene substitutions lead to gross behavioral effects that follow simple Mendelian regularities. Color blindness (J. Bell, 1926), ability to taste phenylthiocarbamide (Snyder, 1932), and some forms of mental retardation are clearly related to simple gene transmissions. Initial observations such as these form the core materials for more complex analyses of the mechanics of gene action.

Typically in behavior genetics studies, traits controlled by several genes are subjected to one of two types of analysis, both aimed at specifying the relative contributions of genes and environment and of providing an estimate of the "number of effective factors" controlling the trait. The most popular approach has been that worked out by Mather (1949) and Broadhurst (1960). It consists of obtaining measures from segregating and nonsegregating breeding populations in an effort to relate genetic to nongenetic variation. From this, the lower limit of the number of effective factors (genetic units) can be obtained. Many behaviors have been studied in this fashion.

The alternative approach is that detailed by Falconer (1960) in which familial regressions are determined for a characteristic within heterogeneous populations. This latter approach is just gaining momentum in behavior genetics. Akin to these techniques is that of selective breeding, where units of genetic material are combined for various degrees of phenotypic expression. The common threads discernible in these procedures are the attempts to demonstrate convincingly a genetic component, determine the degree of genetic influence and environmental plasticity, and set down probability levels for the number of genetic units involved.

Since all of the techniques mentioned are derived from genetic theory and practice, it would appear that behavior geneticists have essentially translated uncritically genetic methodology into the study of behavior as a phenotype. The question remains: Has this been a profitable translation? We think it has, to the degree that the results have demonstrated convincingly that behavior is genetically as well as environmentally controlled and can be genetically as well as environmentally manipulated. However, like some other investigators, we feel that there are theoretical and practical limitations involved. We share the opinion of Caspari (1963) that:

The most simple and apparently the most direct approach to the study of behavior by genetic means will, in my opinion, not turn out to be very promising. I am speaking of the establishment of differences in behavioral characters between strains or breeds of the same species, and the analysis of their genetic behavior in crosses. . . . I cannot think of any meaningful question which would lead us further (pp. 97–98).

It is perhaps true that the polygenic nature of most behaviors has trapped many behavior geneticists at the level of genetic interpretation. Further, few clear physiological pathways are available for study. The effects of individual polygenes are generally small, cumulative, and interchangeable at the biological level. No one set of genes needs to be specific for any one phenotype and, presumably, a number of biological systems could pertain to the same manifest expression. It is extremely unlikely, for example, that the genetic systems and biological superstructures of Tryon's (1940) "bright" and "dull" rats even remotely approach that of Heron's (1941) "bright" and "dull" rats or that of the wide variety of mouse strains that differ in learning ability (Winston, 1964). Phenotypically, the range of variability may exert itself in a similar fashion over a variety of problems; yet the dissimilarity of initial gene pools and the various happenstantial gene arrangements resulting from selection or inbreeding guarantee the existence of diverse physiological pathways. Ordinarily gene analysis must stop at this point.

GENE ACTION AND PHENOTYPE

An alternative to tracing the pathway from the phenotype to the genetic units involved is the approach of tracing the pathway from the gene to the phenotype. Ginsburg (1958) believes that important advances will be made when the individual differences are recognized and used as points of departure rather than endpoints:

I cannot agree with the position taken by a number of colleagues that genetic analysis in behavior needs only to demonstrate that particular kinds of behavior are correlated with particular genetic backgrounds. This may be valuable information for the marriage counselor but hardly for the scientist. In a crude way, we have had this type of knowledge for a long time, and it has not been fruitful. It brings us to the threshold of a locked door. If we are to go further, it must be unlocked with an analytic key, and that key seems to me to be gene action (p. 421).

The usefulness of using gene action as the analytic key presumes a close correspondence between gene action and its directed phenotype. Single-gene systems are more likely to show this property than are polygenes. Several studies can be cited to demonstrate the feasibility of this methodology in selected situations.

Work by W. B. Cotter and Caspari (Caspari, 1960) with the mealmoth, *Ephestia*, emphasizes both the power of a single-gene approach to the study of behavior and the uniqueness of the biological system involved. It is known that the substitution of the allele a for the alternative $a+$ leads to the failure of tryptophan to be oxidized to kynurenin. As a result, pigmentation of several tissues is affected. There is a loss of deep coloration in the adult testes, brain, and skin, and the eyes are red instead of black. One important behavioral effect is that the aa male courts a female more quickly than the $a + a +$ male but is less agile in carrying out the complex series of movements, so that in the long run the wild-type male is more successful in copulating than is the red-eyed male. Here is clear evidence that a single allelic substitution acting through a specific biochemical pathway changes the phenotypic expression of a complex behavior. It is not clear whether this same substitution is similarly effective in altering sexual behavior of other species or even if the usual variation in tryptophan or kynurenin plays a part in the sexual responses of *Ephestia*.

A more commonly known example is that of the human disease, phenylketonuria, which, if untreated in early infancy, results in severe mental retardation. Again, a single recessive substitution in the homozygous condition (pp) is responsible for the metabolic block. Phenylketonurics lack the liver enxyme parahydroxylase, which ordinarily converts phenylalanine to tyrosine. Blood phenylalanine accumulates and may act as a toxic agent to affect intellectual stability. A diet low in phenylalanine, if begun early enough, can ameliorate the condition. Obviously, the knowledge of this circumscribed pathway from the gene to the subsequent phenotype is of great importance in the understanding of this condition, but we wish to emphasize that the results apply only to this one pathway and to this one condition.

Other examples could be cited involving single genes, relatively simple pathways and easily observed behaviors. Of particular note are the several neurological mutant mice that have been studied (for example, Thiessen, 1965b). In general we find that single gene substitutions leading to gross morphologic and behavioral changes are the most easily studied systems. Often the pathways from the gene to the behavior are well marked by major biological events that can readily be traced during their development. At the same time, single-gene effects encapsulate these advantages into unique systems and unfortunately have little to say about polygenic inheritance of behavior such as underlies most behavioral phenotypes. What appears to be needed is a means of specifying physiological functions that have generality beyond the single gene and still show a one-to-one correlation with the observed behavior.

MECHANISM-SPECIFIC BEHAVIOR

As an alternative to the genetic-unit approach that almost inevitably results in excessive overparticularization, we suggest that behavior genetics focus on *mechanism*-specific studies, in which structure-function relationships are delineated such that the identified function is essentially an inevitable characteristic of the identified structure. Although this emphasis has not, to our knowledge, been explicitly related to behavior genetics work, the approach is not new.

Psychology already provides many examples of mechanism-specific analyses. Whether an organism can discriminate different electromagnetic frequencies in the visual spectrum is routinely assessed to determine whether "color" can be utilized for its cue value as a behavioral stimulus. At this level of analysis, the specific mechanism of color vision might in fact be different from one species to another, but the general principle of using electromagnetic differences for their cue value would exist quite independently of the particular genetic or neurological structure per se that mediated the visual capacity. More detailed analyses of the structural mechanisms can, of course, further define the situations over which generality of principles might hold, and sensory psychologists have tended consistently to emphasize such mechanism-specific explorations.

Structure alone, when only limited portions of the total structure are assessed, is not a sufficient basis for establishing the existence of a mechanism. For example, the xiphosuran *Limulus* has peripheral photoreceptors that would provide a basis for color vision but apparently lacks appropriate central nervous system structures to utilize differentially the peripheral cues, and therefore apparently lacks the "mechanism" of color vision (Milne and Milne, 1959).

The assumed complexity of the structures that mediate decision behavior—primarily the central nervous system, but to some extent the endocrine and other systems—has undoubtedly inhibited psychology from assuming responsibility for structure-function analyses. Nevertheless, increasingly precise techniques are becoming available for just such analyses. Differences in brain biochemistry (for example, Krech, Rosenzweig, and Bennett, 1960), specific brain localizations (for example, the classical work of Olds and Milner, 1954, and subsequent work in the same tradition), much work of the sensory physiologists in mapping neurological processes of a variety of sorts (for example, Hubel, 1963, and related work), and increasing work on endocrine function and behavior (the work of Beach and his group on sexual behavior and gonadal hormones, Beach, 1964, and the work of Young and his group on endocrine control of both morphology and behavior, Young, Goy, and Phoenix, 1964, among others) are increasingly demonstrating our capacity to make structure-function analyses and to relate behavioral outcomes to specific mediating mechanisms.

The techniques of behavior genetics are especially adapted to structure-function studies. Genetic differences can outline the extent of mechanism-specific behaviors; crosses can verify structure-function relations; and selection techniques can set predetermined levels of biological activity. For example, Rosenzweig (1964) and his associates utilize genetic material to identify and manipulate relatively small changes in brain chemistry associated with various kinds of experiences, changes that could be obscured in biologically variable material. McGill has shown the utility of genetic control in elucidating the sexual response of the mouse (McGill and Tucker, 1964). In our own work on alcohol preference (for example, Rodgers and McClearn, 1962; Rodgers, 1966), genetic control has been indispensable in stabilizing metabolic processes that appear to be crucially involved in the alcohol-drinking behavior of the mouse. These are but a few of

a growing series of studies that make use of genetic controls more for standardization and alteration of the behavioral unit than for identification of the referent
population. Moreover, biological modification is done without physical insult to
the animal and without undue concern for genetic details.

Once the mechanistic relationships are assessed, they can be generalized to all
situations in which similar relationships are to be found. They tend to provide
convincing "explanations" of the observed phenomena. For example, a genetic
abnormality can give rise to cretinism, a condition that results in profound mental
impairment of the developing child. In the genetically stabilized condition, the
impairment has been traced to the absence or inadequate functioning of the thyroid. Mental impairment also occurs in rats (as well as other species) following
removal of the thyroid at birth. The impairment in rats has in turn been related
to a lack of proliferation of dendritic structures in the developing brains. The
condition is irreversible if deficiency occurs during the period of dendritic proliferation but is reversible if it occurs subsequent to the development of these
neural structures (Eayrs, 1959).

Similarly in humans, thyroid insufficiency early in life, whether induced genetically or otherwise, results in irreversible mental changes. Later in life, it gives
rise to the largely reversible condition of myxedema. The results are thus highly
suggestive—and where still in doubt are subject to specific test in feasible studies—
that directly parallel effects occur in rats and in humans, that thyroid insufficiency
during a critical developmental period is a more uniquely causative agent than
is any kind of genetic constitution, and that insufficient dendritic development
in the central nervous system is at least one mechanism through which the behavioral effects are mediated. If correct, the "explanation" that the mental deficiency of cretinism is a consequence of the failure of development of specific
central nervous system structures carries a degree of self-evident conviction. It
should be generalizable to any situation in which similar structural defect can
be identified. It thus becomes a rather complex although relatively precise mechanism-specific explanation, growing in part out of the study of genetically
stabilized material but not uniquely characteristic of or conceptually limited to
a given genotype.

Analyzing the relationships in mechanism-specific terms suggests additional
clarifying research projects. For example, clarification of how thyroid output
stimulates or is necessary for dendritic proliferation is an obvious next step,
and it might clarify many characteristics of the cretin syndrome. Conceivably,
the resulting discoveries might allow future selective stimulation and impairment of neurological development in experimental animals, with resulting precision in the study of structural-functional neural relationships. Inevitably, such
structure-function studies identify functional subunits that have potential utility
in a variety of situations remote from the particular ones in which they were
isolated. In contrast, specific genetic material studies and broadly general laws
that do not relate to specific structures usually lack the quality of transposability
to other situations.

Cancer research and other disease studies have long made use of genetically stabilized material to explore possible mechanism-specific relationships. Most of the inbred strains of mice currently available were developed explicitly for such purposes. Our proposal is to extend such mechanism-specific concerns to the study of behavior, utilizing the techniques of genetic control to stabilize behavior segments and the related structural mechanisms. Behavior genetics would be an appropriate specialty within which to undertake such studies. Conversely, attention to the isolation and identification of mechanism-specific relationships would rescue behavior genetics from the alternative pitfalls of overparticularized studies of ungeneralizable phenomena and overgeneralized studies of phenotypes treated independently of structure-function considerations.

To summarize, we argue for the use of behavior genetics as a tool to standardize structure-function relationships so that mechanisms underlying different behavioral processes can be isolated and precisely identified. We specifically question the use of behavior genetics as a device for identifying genetically homogeneous groups across which generalizations can be made. In our conception, identification of strain differences and demonstration of genetic manipulability of a behavioral phenotype is simply the initial demonstration to allow further study of the mechanisms underlying the phenotype. Identification of strain differences is not an end in itself. We similarly question the utility of ignoring mediating mechanisms, in input-output analyses based on the "black box" fiction. It is our contention that only mechanism-specific identities provide a parsimonious basis for generalizing from one behaving organism to another and that the methodologies of behavior genetics are uniquely suitable for such analyses.

BIBLIOGRAPHY

BEACH, F. A.
1960. Experimental investigations of species-specific behavior. *Amer. Psychol.*, 15:1–18.
1964. Biological bases for reproductive behavior. In W. Etkin (Ed.), *Social Behavior and organization among vertebrates*. Chicago: University of Chicago Press.
BELL, J.
1926. Color-blindness. *Treasury of human inheritance*, 2:125–267.
BELL, R. Q.
1965. Developmental psychology. In P. R. Farnsworth (Ed.), *Annual review of psychology*, 16:1–38.
BITTERMAN, M. E.
1965. Phyletic differences in learning. *Amer. Psychol.*, 20:396–410.
BROADHURST, P. L.
1960. Applications of biometrical genetics to the inheritance of behavior. In H. J. Eysenck (Ed.), *Experiments in personality*. London: Routledge and Kegan Paul.

BRUELL, J. H.
1962. Dominance and segregation in the inheritance of quantitative behavior in mice. In E. L. Bliss (Ed.), *Roots of behavior*. New York: Harper.

CASPARI, E. W.
1960. Genic control of development. *Perspectives Biol. Med.*, 4:26–39.
1963. Genes and the study of behavior. (Vice-presidential address.) *Amer. Zoologist*, 3:97–100.

COLLINS, R. L.
1964. Inheritance of avoidance conditioning in mice: a diallel study. *Science*, 143:1188–1189.

DE FRIES, J. C.
1964. Prenatal maternal stress in mice. *J. Hered.*, 4:289–295.

DE FRIES, J. C., M. W. WEIR, and J. P. HEGMANN
1965. Blocking of pregnancy in mice as a function of stress: supplementary note. *Psychol. Rep.*, 17:96–98.

EAYRS, J. T.
1959. The status of the thyroid gland in relation to the development of the nervous system. *Anim. Behav.*, 7:1–17.

FALCONER, D. S.
1960. *Introduction to quantitative genetics*. New York: Ronald Press.

FULLER, J. L., and W. R. THOMPSON
1960. *Behavior genetics*. New York: John Wiley.

GINSBURG, B. E.
1958. Genetics as a tool in the study of behavior. *Perspectives Biol. Med.*, 1:397–424.

HERON, W. T.
1941. The inheritance of brightness and dullness in maze learning ability in the rat. *J. genet. Psychol.*, 59:41–49.

HIRSCH, J.
1958. Recent developments in behavior genetics and differential psychology. *Dis. Nerv. Syst., Monogr. Suppl.*, 19:1–7.
1962. Individual differences in behavior and their genetic basis. In E. L. Bliss (Ed.), *Roots of behavior*. New York: Harper.

HUBEL, D. A.
1963. The visual cortex of the brain. *Sci. Amer.*, 209:54–62.

JAY, G. E.
1963. Genetic strains and stocks. In W. J. Burdette (Ed.), *Methodology in mammalian genetics*. San Francisco: Holden-Day.

KRECH, D., M. R. ROSENZWEIG, and E. L. BENNETT
1960. Effects of environmental complexity and training on brain chemistry. *J. comp. physiol. Psychol.*, 53:509–519.

LINDZEY, G., H. D. WINSTON, and M. MANOSEVITZ
1963. Early experience, genotype and temperament in *Mus Musculus. J. comp. Physiol. Psychol.*, 56:622–629.

McCLEARN, G. E.
1960. Strain differences in activity in mice: influence of illumination. *J. comp. physiol. Psychol.*, 53:142–143.

McCLEARN, G. E., and W. MEREDITH
1966. Behavior genetics. In P. R. Farnsworth (Ed.), *Annual review of psychology*, 17:515–550.

McGILL, T. E., and G. R. TUCKER
1964. Genotype and sex drive in intact and in castrated male mice. *Science,* 145: 514–515.

MATHER, K.
1949. *Biometrical genetics.* New York: Dover.

MEISSNER, W. W.
1965. Functional and adaptive aspects of cellular regulatory mechanisms. *Psychol. Bull.,* 64:206–216.

MERRELL, D. J.
1965. Methodology in behavior genetics. *J. Hered.,* 56:263–266.

MILNE, L. J., and M. MILNE
1959. Photosensitivity in invertebrates. In J. Field, H. W. Magoun, and V. E. Hall (Eds.), *Handbook of physiology. Section 1: Neurophysiology.* Vol. I. Washington, D.C.: American Physiological Society.

OLDS, J., and P. MILNER
1954. Positive reinforcement produced by electrical stimulation of septal area and other regions of rat brain. *J. comp. physiol. Psychol.,* 47:419–427.

RODGERS, D. A.
1966. Factors underlying differences in alcohol preference among inbred strains of mice. *Psychosomatic Med.,* 28:498–513.

RODGERS, D. A., and G. E. McCLEARN
1962. Alcohol preference of mice. In E. L. Bliss (Ed.), *Roots of behavior.* New York: Harper.

ROSENZWEIG, M. R.
1964. Effects of heredity and environment on brain chemistry, brain anatomy, and learning ability in the rat. *Univ. Kan. Symp., Kansas Studies Ed.,* 14:3–34.

SCOTT, J. P., and J. L. FULLER
1963. Behavioral differences. In W. J. Burdette (Ed.), *Methodology in mammalian genetics.* San Francisco: Holden-Day.

SNYDER, L. H.
1932. The inheritance of taste deficiency in man. *Ohio. J. Sci.,* 32:436–440.

THIESSEN, D. D.
1965a. Persistent genotype differences in mouse activity under unusually varied circumstances. *Psychonomic Sci.,* 3:1–2.

1965b. The wabbler-lethal mouse: a study in development. *Anim. Behav.,* 13:87–100.

THIESSEN, D. D., J. F. ZOLMAN, and D. A. RODGERS
1962. Relation between adrenal weight, brain cholinesterase activity, and hole-in-wall behavior of mice under different living conditions. *J. comp. physiol. Psychol.,* 55:186–190.

TRYON, R. C.
1940. Genetic differences in maze-learning ability in rats. *Yearb. Nat. Soc. Stud. Educ.,* 39 (1):111–119.

WADDINGTON, C. H.
1957. *The strategy of the genes.* London: George Allen and Unwin.

WINSTON, H. D.
1964. Heterosis and learning in the mouse. *J. comp. physiol. Psychol.,* 57:279–283.

YOUNG, W. C., R. W. GOY, and C. H. PHOENIX
1964. Hormones and sexual behavior. *Science,* 143:212–217.

SENSE PERCEPTION AND BEHAVIOR

HANS KALMUS

T HE THEME assigned to me—the behavioral consequences of genetically determined differences in sense perception—is so vast that I can only partially cover my field. Leaving out the more obvious aspects of blindness and deafness I propose to deal only with some less known consequences of these conditions and to discuss more fully some lesser defects of the eye, the ear, and the chemical senses, although I cannot do even this fully or systematically. I shall also touch on the taste deficiency for PTC, which, while common in our population, is quite rare elsewhere.

Most situations in life are greatly complex and to some extent unique. They cannot be systematically repeated nor do they lend themselves to easy experimental variation or statistical analysis. This also applies to the behavioral responses to these situations. Incidents occurring in ordinary life outside the laboratory, which lead to the discovery of deficient sense perception, are thus very often anecdotal, though in retrospect they can be classified according to a finite number of impaired skills. Two famous cases, both described in a paper by Chapanis (1951), may serve as examples.

A collision occurred on the night of July 5, 1875, between the steamship *Isaac Bell* and a tug off Norfolk, Virginia. The two vessels crashed and ten lives were lost. During the investigation the master and officers of the steamer testified that the tug captain, at the relevant moment, could have seen only the steamship's green starboard light, but the captain of the tug insisted that he had seen a red light and had altered his course accordingly. The incident was only cleared up four years later when a surgeon, checking on the tug captain's vision, discovered that he was so color blind he could scarcely distinguish between a red and a green light at a distance of three feet.

Another well-documented and somewhat capricious incident concerns the reputation of the Jesuit priest Maximilian Hell, a leading astronomer of the eighteenth century. In 1769, this scholar had made some observations of the transit of Venus, recording them in a journal later deposited in the Vienna Observatory. Sixty-four years later, Joseph Johann von Littrow, then director of this observatory, announced a scandal that upset the astronomical world: he thought he had found clear evidence that many of Hell's entries had been altered. He contended that many original notations had been scraped out and replaced by new writing in ink of an entirely different color, thereby concluding that Hell had faked some of his data. In 1883, however, the American astronomer

73

Simon Newcomb had another look at Hell's journal and discovered that the differences in ink which Littrow had described were not actually color differences but merely variations in intensity of the same color. Under a magnifying glass, it also appeared that no erasure had ever been made and that Hell had simply traced over his original writing with a darker ink. Newcomb suspected Littrow had been color blind, and indeed he found out from some of the observatory's staff that Littrow had been unable to distinguish between the red color of aldebaran and that of the whitest star. Thus not only was the reputation of Hell restored but an important piece of work, which by innocent but mistaken testimony of a color-blind man had been made suspect, was reinstated into its validity.

Special instances of the behavioral effects of genetical sensory defects in man can be better understood in relation with other fields of biological inquiry, such as comparative sense physiology, the study of phenocopies, experimental psychology, the study of group interactions, and problems of cultural anthropology. The importance of the sense organs for the faculties of man and animals was first emphasized by von Uexküll (1921) who showed that the analytical power of a sense organ is a limiting factor in the perception of an animal, so that one cannot expect any animal to react, for instance, to visual details beyond the resolving power of its eyes.

I have shown that in a variety of insect species white-eyed mutants, who lack the pigments that optically isolate the omnatidia of their compound eyes, cannot perceive contour movement (Kalmus, 1943, 1961). In nature such individuals are hardly viable, though a white-eyed colony of the American cockroach seems to thrive in a Welsh coal mine (Jefferson, 1958), but in the laboratory they can be mostly kept. However, *Drosophila subobscura*, unlike other species of this genus, must see each other in order to court and to mate. Normal cultures, therefore, cannot be kept in the dark and the white-eyed mutant cannot be kept at all, even in the light (Philip *et al.*, 1944).

The genetically determined sexual dimorphism of the vast majority of animal species provides many examples of differences in the sensory equipment and of special releasers characteristic for one sex or another. For instance, the odor glands at the wing base of many female moths and the correspondingly specialised antennae of the males together play an important role in the sexual behavior of the species concerned. Sometimes different complex scent organs and correspondingly difficult chemoreceptors recur in both sexes (Barth, 1952). Similar sexual differences are also highly developed in other insects, for instance, in the hymenoptera. The differences of the various castes of honey bees in their visual and olfactory equipment (Kalmus, 1960), as well as of their odor production, are a continuing source of interesting discoveries (Butler, 1964). The behavior, and in particular the sexual behavior, of sensory dimorphic species provides many instances on which to model our ideas concerning the behavior of sense-defective people.

Also facilitating the study of hereditary sense deprivation is the existence of numerous phenocopies, acquired sensory defects, that are caused by disease and accidents and, occasionally, even by therapeutic measures. These provide information concerning the effects of the time of onset on the behavioral consequences of deafness, blindness, and other conditions.

An interesting—and rare—example is the loss of vestibular function in a doctor after prolonged streptomycin treatment for tuberculosis. The patient described how he gradually overcame the most alarming sensations of vertigo, nausea, and his inability to move or even to read, and how, despite the fact that his vestibular function remained impaired, he learned with the help of his other sense organs to live an almost normal life as a clinical research worker (Crawford, 1952). Reports of how the much more frequent handicaps of hereditary and acquired blindness and deafness are behaviorally compensated abound, but a modern systematic study might be very interesting.

The story of Helen Keller (1927) and other less famous examples show how even the joint handicap of early acquired blindness and deafness can be overcome to an extraordinary degree. The extent of adaptation and compensation which, by great skill and personal effort, could be achieved in cases of hereditary defects is as yet unknown.

For experimental purposes one can simulate to a degree a number of sensory deficiencies. Some of the means are simple and obvious like confining people for short periods to the dark, blindfolding them, or using monochromatic light; blocking people's ears and pinching their nose; or using local anesthetics. Other more subtle means are perpetually being developed. Changes in sensory perception of a large or permanent nature result sometimes from operations (for example, juvenile cataract) or they can be produced in anthropoids.

Genes responsible for defective sense perception have other manifestations as well; and in certain circumstances it is not obvious that the sensory manifestation of such a gene is the important one behaviorally or with respect to selection. This may apply in the case of the PTC-taster polymorphism, where the food preferences of the various genotypes may differ, on the one hand (Fischer *et al.*, 1961), but where, on the other hand, taste differences in actual food are too small to account for any choice reaction. However, the thyroid conditions that are in some ways associated with the PTC phenotypes certainly have clinical and thus also behavioral effects (Kalmus, 1964). Sensory deficiencies are components of many hereditary diseases and syndromes, which clinically may sometimes be dominated by other features. Defective vision, scotomas, partial deafness, and vestibular-cerebellar disturbances may for instance be early recognizable symptoms of retinitis pigmentosa (Hallgren, 1959), while some degree of mental deficiency may also occur with the other symptoms. The behavior of the individuals suffering from this and similarly complex syndromes will considerably vary in their behavioral responses, dependent on the onset or the predominance of a particular pleiotropic manifestation of the gene. Sensory defects provide good reasons for disqualifying candidates who want to enter certain professions and occupations, though not always absolutely compelling reasons. To understand this puzzling practical situation it is necessary to ask the fundamental question of what a particular sensory defect implies in the way of defective behavior and performance and what it does not imply. It has often been naively neglected and conversely sometimes been overstressed that the vast majority of human activities and skills are monitored not by one sense modality alone, but by several, and that when one of these sensory feedback loops is defective or even absent, others will be more strongly developed so as to compensate, albeit to varying degrees, for the deficiency. In this context I might perhaps mention that hereditary color blindness, which in its various forms affects about 8 per cent of all males in European populations, was not recognized until the eighteenth century.

The earliest case on record seems to be that of Harris, the shoemaker, reported in 1777 by Huddart in a letter to the Rev. J. Priestley. Huddart wrote:

The account he [Harris] gave was this: that he had reason to believe other persons saw something in objects which he could not see; that their language seemed to mark qualities with confidence and precision which he could only guess at with hesitation, and frequently with error. His first suspicion of this arose when he was about four years old. Having by accident found in the street a child's stocking, he carried it to a neighbouring house to inquire for the owner. He observed the people called it a red stocking, though he did not understand why they gave it that denomination, as he himself thought it completely described by being called a stocking.

From this description it is fairly clear that Harris was suffering from protanopia. Since his time a vast number of color defectives have been described and the existence of several other familial forms established (see Kalmus, 1965). There is little doubt that these strictly hereditary forms of sex-linked color deficiency have existed since the dawn of history, and yet, among the hundreds of Renaissance painters and their thousands of apprentices, not one single case has been spotted or at least found worth mentioning. Not even a genius like Leonardo observed color blindness. We must thus conclude that it needs a particular mental and social climate for certain specific defects to become sufficiently obvious and have behavioral significance.[1] In our case it must have been the requirements of textile colorists, traffic signals, and scientific activities that converted color blindness from a hidden into an overt phenomenon.

Changes in the social climate and scientific advances have completely changed the social status and thus the behavioral pattern of the people suffering severe sense deprivation like total blindness or deafness. Such individuals in olden times were mostly condemned to live a marginal existence, sometimes almost outside society, thus developing all sorts of antisocial behavior. They were later kept in institutions, but they are nowadays, by the modern means of education and technology, largely integrated into the main social body from which they stem and show a behavior much nearer the norm.

An important factor determining the behavioral consequences of sense dep-

---

1. Vernacular words for blind and deaf occur in most developed languages, but are lacking for the ageusias and anosmias. It seems likely that a lack of the faculties of smell or taste might have been very deleterious in primitive conditions, but language was probably not advanced enough to describe these conditions. Of course, the names may also have been lost.

rivation is the onset of the condition. An early onset, and in particular the congenital presence of total defect, makes certain primary experiences like those of sound or color impossible and alters the development of an individual's mental faculties in various specific ways. The congenitally blind and the congenitally deaf learn in later life how to use the words for color, brightness, music, or noise, but these remain "empty" words. These same congenitally blind or deaf learn early to use their unimpaired faculties in a special way and thus perform many tasks in a way specifically different from the normal. On the other hand, late and insidious onset of conditions like otosclerosis or cataract does not destroy the memory of previous sensory experiences, and it modifies or hampers many well-developed adult faculties only to a lesser extent. Some effects of the insidious developments of visual and auditory defects on painters and composers will be discussed below.

The intensity and sometimes even the kind of a behavioral response to a sensory deficiency depends also on the degree of this deficiency. Defects of vision are to a certain degree compensated for by habits, by corrective lenses, or by improved illumination, which may all be used jointly in special schools for children with impaired vision. But below certain levels of visual faculties these means have to be abandoned and the quite different means of nonvisual education of the schools for the blind must be used. This situation has a curious parallel on the eye development of animals living in the dark, who may either develop enormous eyes or sometimes lose them altogether.

GROUP EFFECTS

The effects of hereditary sense deprivations are not confined to their carriers but may extend to the various biological and social groups of which they are members, that is, to their mates, families, school fellows, crews, passengers, and even the public at large. When these potential effects are dangerous, measures are usually taken to prevent them from materializing. In the case of color blindness for instance, it is possible to exclude all dichromats and, according to the situation, also many severely anomalous trichromats from certain tasks in signaling and other technical branches of the armed forces, as well as in public transport.

However, it is impossible to exclude very large numbers of people from driving private automobiles. Up to one twelfth of all potential male drivers may be color blind and thus a remedy must be sought which does not make it impossible for these men to drive. One obvious way is a modification of existing signals and in particular of traffic lights. One can, for instance, standardize the position of the lights or alter their spectral composition. In Great Britain the red light is always on top of a traffic signal, the amber light in beween, and the green light at the bottom, so that in many situations even a color-blind driver can read the signal. I am also told, although I cannot trace the source, that in the same country the composition of the green light, which had originally been a more yellowish green, has been changed to a bluish green with a view to making it more distinct from the red. I must say I cannot quite see how this would work for a protanope. Dr. Sloan, of Johns Hopkins Hospital, Baltimore, was reported by Pickford (1955) to be working on signal lights which might be satisfactory for color-vision-defective subjects without being much less than optimal for normals.

The necessity and urgency of such measures is at present difficult to assess as to my knowledge no statistics exist on the role defective color vision may play in road accidents. In the absence of such information it is also difficult to assess whether the automobile is at present replacing those selective agencies which in the past may have acted against the spread of the mutant genes responsible for color vision deficiency. Even if it could be shown that a certain proportion of road accidents is caused by color-blind people, as we all know, it is not always the color blind involved who is the victim of his deficiency, but frequently others. Thus the efficacy of road accidents as selective forces against the color vision genes might be very slight indeed.

The importance of color blindness in the textile industry (Richter, 1953) and the printing trade (Lakowski, 1962; Pickford, 1955) have been made the subject of special study.

Diminished sensory discrimination sometimes affects diagnosis and therapy not only on the patient's side, but also on the part of the doctor. The importance of any particular deficiency depends largely on the doctor's speciality. The eyesight of a surgeon ought to be perfect or at least its defects compensated, while this is less important in a psychiatrist. Deafness is less dangerous for a doctor since, as a result of X rays, auscultation and percussion have greatly decreased in importance. A color-blind doctor, especially a protanope, may have difficulty in recognizing cyanosis or the peculiar color of the lips and the blood in carbon-monoxide poisoning, especially in some types of artificial illumination. There probably are very many other occasions where diminished sense perception constitutes a hazard for patients, but a general remedy for this would be difficult to devise.

SENSORY DEFECTS AND ART

A great deal of nonsense has been said and printed concerning the effects of sensory defects on artists and works of art. Nevertheless, much of interest has also been reported. The powerful effects of otosclerosis on the playing of aging violinists and other virtuosos suffering from hereditary hearing defects have been perceived by many music lovers, while the tragic gradual loss of hearing on composers such as Beethoven or Dvorak has been well reported by these artists' biographers.

Rather more humorous is the frequent familial condition of tune deafness (Kalmus, 1948), a congenital inability to perceive, memorize, or reproduce melodies that runs in families, though the precise mode of its inheritance is not clear. Children singing everything in monotone or to the same "tune," generals not recognizing their national anthem, and piano students not noticing their most frequent mistakes are very striking examples of the effects of tune deafness. A small percentage of most European populations are tune deaf; these should not be encouraged to produce music. On the other hand, they often enjoy hearing it.

In the visual arts, many attempts have been made to interpret certain peculiarities of technique and presentation as due to specific hereditary deficiencies be they of refraction, acuity, or color sense. Trevor-Roper (1957), to whom most of the ensuing discussion is due, has rightly pointed out that such simple "physiological" explanation of pictorial style and peculiarities can only be fairly applied to naturalistic paintings, which in most periods and civilizations have been of little importance.

The most famous examples are the paintings of El Greco, most of which show a slightly oblique elongation of his characters. This has frequently been ascribed to astigmatism, but apart from the fact that one cannot easily see how this could work, it appears from X-ray photographs that perfectly proportional outlines were sketched under the elongated final versions. So it is more reasonable to ascribe El Greco's manner to a Cretan compromise between Venetian naturalism and Byzantine stylization.

On the other hand, it is very likely that myopia contributed to the peculiarity of impressionistic painting: the use of blurred masses without much detail. Many impressionist painters—such as Degas, Braque, and Matisse—were myopic, and Renoir, who also suffered from this disability, is known not to have worn glasses. Presbyopia may also have comparable effects on painting.

The astonishing fact that color blindness has not been reported prior to 1777 has already been mentioned. No traces of it can be found in Renaissance paintings. However, the pictures of several painters at the London exhibition of 1872 showed "the roof tops" and "the oxen" red on the lit side and green in the shade, which one might interpret as a sign that the painters were color blind. Attempts at explaining certain color mannerisms of the modern schools as a consequence of defective color vision are quite frequent but only rarely convincing, since most workers do not attempt a naturalistic rendering of anything in particular. A color-blind painter may occasionally use his defect, making a virtue out of necessity, either by using strong "primitive" colors, which he can perceive, or by not attempting to differentiate between hues, which he himself cannot distinguish.

The effects on paintings of cataract—and its removal—have also been demonstrated in several instances.

Leaving this rather subjective field, we will now turn to more tangible biological consequences of sense deprivation, namely, the choice of mates and its genetical effects.

PHENOTYPICAL ASSORTATIVE MATING

A very striking difference exists between the blind and the deaf concerning their choice of mates. In general deaf people are often found to marry other deaf people, while the union of two blind people is exceedingly rare. The reasons for this difference seem obvious enough and can be directly traced to the different roles vision and hearing play in an individual's behavior. As pointed out earlier, vision serves primarily to inform us about the position and movement of objects in our physical environment, including other people, but it does not normally contribute much in the way of conversation. The main function of hearing, on the other hand, is just the communication with other individuals, while acoustic localization is of lesser importance. The deprivations of the blind and the deaf are therefore very different during infancy and remain different, in spite of many mental adaptations and modern artificial aids.

This also results in differences in the requirement such people have concerning their mates. The spatial disorientation of a blind boy cannot be much improved by cooperating with a blind girl and vice versa, as both partners would on many occasions be equally helpless. To a blind person another blind person does not appear particularly attractive or different from a seeing person, except that he is clumsy and unable to take a lead in vital situations where a seeing partner naturally will do so. Mental irritation is therefore quite probable. On the other hand, to a seeing person, the very helplessness or alternatively resourcefulness of a blind person of the opposite sex may be very attractive and conducive to an easy establishment of cooperation. Thus blind people usually pair up with seeing partners.

The situation of deaf people is quite different. Their main means of spatial orientation—vision and touch—are not impaired, but although they can see other people, they cannot easily communicate with them and are mostly excluded from their conversations. By the time two deaf individuals reach maturity, they have usually developed some common means of communication, frequently the well-known form of elaborate sign language. A deaf person, therefore, is frequently better able to satisfy the social needs of another deaf person than a hearing person is. On the other hand, speaking to the deaf is always a strain, and some characteristics, such as the tendencies to pretend being able to hear and to be suspicious, often found in deaf people seem to be less endearing to normal people than are characteristics of the blind. All this adds up to a very strong tendency for the deaf to marry among themselves, as Figure 1 indicates.

It should be mentioned in passing that these differences between the mating habits of the blind and the deaf have important consequences for the student of the genetics of the several defects affecting the eye and the ear. On the one hand, the negative assortative mating of the blind makes the study of autosomal recessives difficult and thus focuses attention on dominant or sex-linked conditions, while on the other hand, the great propensity of deaf to intermarry has made the recessives much more prominent in the picture. It also has provided the human geneticist with the rare opportunity of distinguishing between naturally occurring cases of allelism and nonallelism, as Figure 2 shows (Hopkins and Guilder, 1949; Hanhart and Luchsinger, 1951).

The assortative mating of the deaf that frequently results in deaf offspring is often a consequence of the coeducation of deaf children and of their meeting in clinics or social clubs. By educating deaf boys and deaf girls separately and by

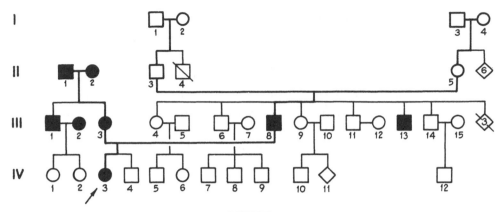

FIGURE 1

Pedigree of deaf people (from Hopkins and Guilder, 1949). The deafness of II2 was erroneously attributed to scarlet fever, whereas III1 and 3 show that it must have been hereditary, probably recessive. IV1 and 2 have good hearing and show that the deafness of III1 and 2 was not caused by the same recessive gene; though there was a history of family deafness in III2, there was, as well, whooping cough and measles at age 4. III3 and 8 probably were homozygous for the same recessive. The arrow indicates the proposita.

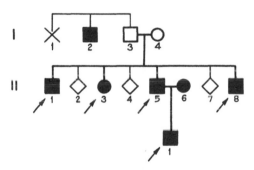

FIGURE 2

Pedigree of deaf people (from Hopkins and Guilder, 1949). The arrows indicate that five members of this family were at various times pupils of the same deaf school. The cause of deafness in II6 is unknown.

reorganizing deaf clinics and clubs, one could undoubtedly reduce the incidence of deaf (mainly deaf-mute) children. One must, however, try to balance any such measure against the benefits accruing from the improved personal relations among deaf couples and families.

TYPES OF BEHAVIOR

A systematic treatment of our material could be attempted in several ways. Sensory deficiency becomes manifest only in certain situations, which it may be useful to survey.

Behavior toward Nonhuman Objects

In spite of such interesting Utopias as H. G. Well's Country of the Blind, it is unlikely that a human society entirely composed of blind people would be viable, and to my knowledge there is no report that such a community ever existed. Nor, as a matter of fact, are there any reports of a self-supporting community of deaf people, although such a group might perhaps be more able to exist in certain circumstances.

Concerning the minor deficiencies, we might think that the impoverishment of common perception and the resulting loss of discrimination must in many ways reduce efficiency. All forms of defective color vision will make food gatherers and hunters less efficient in the finding and choice of food or game and may, on the other hand, also make them prone to overlook predators, thus succumbing to attacks they would have otherwise survived. These aspects have been discussed by Post (1962).

A visual defect that attracted a good deal of attention during World War II is night blindness. This condition is sometimes caused by vitamin A deficiency, and these forms can be successfully treated by supplying this substance. A minority of cases, however, are stationary and hereditary, presumably owing to an abnormal retinene cycle, and are hardly affected by treatment. The mode of inheritance is either dominant autosomal or sex-linked recessive (Riddell, 1940). People suffering from stationary night blindness are inordinately helpless when suddenly stepping into the (incomplete) dark, and some types show no adaptive improvement even after prolonged periods in low illumination. Ordinarily the condition does not cause too much inconvenience, but it was frequently noted by civilians during the blackout and by servicemen on night patrol or marches. It certainly makes a person unfit for combat duties. Simulating night blindness, unlike the simulation of color blindness, is very difficult. At earlier stages of anthropogenesis, it must have been a considerable handicap.

Detrimental effects must also be expected from minor auditory deficiencies. Monaural deafness, for instance, seriously impedes acoustic localization; hardness of hearing again is particularly dangerous in the dark. Ageusias may prevent the rejection of decomposed food, and so can anosmias. The latter also have some importance in industrial societies where, for example, people who cannot smell coal gas are more likely to be poisoned by carbon monoxide than people having the full possession of their smelling powers.

It is not customary to dwell on the (lethal) behavior of people who perish because of some inborn sensory deficiencies; thus we know little about the reactions of most such unfortunate individuals prior to their death. While this is understandable from an esthetic and sentimental point of view, it has the unfortunate consequence that this kind of tragic pathology is sometimes not prevented.

The development of some basic components of the mental apparatus is not completely inhibited by sense deprivation, though it is often greatly impeded. This applies to many concepts concerning time and space. Many of these—like the ideas of earlier and later, of nearer, higher, larger, and so on—can be inferred from tactile, visual, or auditory cues and normally are inferred from all three; in the absence of one, these ideas can develop with the help of two of the others. Helen Keller is an example of how even an early loss of both sight and hearing need not result in an atrophy of such concepts. On the contrary, she acquired these concepts, as well as the rudiments of mathematics, a high degree of literacy, and a firm ethical system, after the loss of both main senses.

Social Behavior

The role of sensory perceptions in the social behavior of any person is enormous and of great variety. They mediate the recognition and classification of other people and are instrumental for liking or disliking particular individuals. Losses in a sensory faculty obviously must greatly affect such attitudes as well as all integrated social activities. More than in spatial-temporal orientation, a comparative study of the role of sensory diversity among the sexes and castes of nonhuman social organisms would be revealing here. Some examples have been mentioned on page 74. The following scheme might help in the consideration of the effects of sensory deficiency on human relations.

1. *Sexual relations.* Normally attraction and repulsion are mediated by the appearance, voice, and odor of a partner. The inability to perceive any of these affects the choice of a partner and the relations with him (see also pages 79–80).

2. *Child-parent relations.* These are also modified by the absence of sensory information.

3. *Relations between siblings.* Sense deprivation usually affects the hierarchical structure in a family.

4. *General family relations.* Recessive defects can alter the attitude to consanguineous persons of the opposite sex.

5. *Relations in age groups.* Sense deprivation often produces backwardness and the exclusion of affected children from normal children of the same age.

6. *Relations in employment.* Defective perception can be a reason for excluding a person from certain occupations but also occasionally conducive to others: myopia is said to be advantageous for work involving fine visual detail.

7. *Other pursuits.* Certain sports and other leisure occupations may be more dangerous or less enjoyable.

Behavioral Defects concerning Communication

Language comprises some of the most important forms of human behavior like speaking, listening, writing, and reading. Any of these and other related activities can be profoundly affected by sensory defect or peculiarity. The inability of the congenitally deaf to understand and to develop speech unaided are as well known as the inability of the blind to read or to learn to write without special help. So are the methods of lipreading, sign language, tactile sound instruction, braille, and "moon" (for people who lost sight after having learned to read), as well as modern electronic methods.

It is much less realized, however, that even in the normal range the roles played by the various sense perceptions are exceedingly variable (Galton, 1907), and that similar faculties—for example, counting—in one person may be entirely dominated by vision, while in another the ear is dominating, and in a third the proprioception of an activity. None of this may show in a person's vocabulary. There is little doubt that some of the minor sense deficiencies must be quite decisive in such alternative developments. But as yet little is known in this respect.

We know, however, that language enters into the problems of minor sensory defects in many subtle ways. It is useful to regard the verbal classification of colors as an attempt at dividing a perceptual continuum (the color continuum) into separate and discrete entities, each characterized by a basic color name. These categories are limited in number, usually to between three and eight, and decide to a considerable extent an individual's reactions in situations where color is important, mainly by way of the recognition of objects. But the main importance of color names is that they serve as means of communication between individuals. Ray (1953) has shown that the normal members of different Amerindian tribes divide the spectrum verbally in many strikingly different ways, which are characteristic for each civilization and different from the way in which white North Americans make their division. These color name systems have nothing to do with the human eye's physiological limitations, and Ray could not explain them by any specific racial or linguistic factor.

Members of any particular civilization who are deficient in their color perception, the congenitally color blind, for instance, have to fit a reduced number of basic sensations to the full number of color names in their language and, using secondary clues, often do so successfully.

Verbal methods of discovering defective color vision are uncertain and sometimes deceptive, even when applied to individuals of one's own civilization. It may very well be that persons having an alien color vocabulary and thus an alien subjective color system may show effects of this even when doing nonverbal color vision tests.

It takes a high degree of sophistication and objectivity for color-blind individuals to realize that other people can perceive qualities they cannot themselves see, as the description of protanopia, to which reference has already been made, indicated. Most such people, having assigned color names to certain hues, shades, and objects, will tenaciously stick to these names. Thus, I found a high proportion of young students, even science students, quite unaware that they were color blind. Another remarkable case was a colleague, now a highly regarded philosophy professor, who, on being told that my tests clearly showed him to be a protanope, hotly denied that there was anything wrong with his color vision. An even more ludicrous situation arose when a fellow geneticist, testing pedigrees for sex linkage, for a considerable time did not realize that he was himself color blind and that his classification of many people was rather peculiar.

It is perhaps somewhat easier for the color-normal investigator to detect and to visualize other people's deficiency, but not much. For example, an instructive film, made by the U.S. Navy Department, was produced with the aid of a woman color blind in one eye only.

On several occasions, young couples came to my laboratory to be tested for color blindness, and the boy most frequently declared the girl to be color blind. Invariably it turned out the other way around.

The problem of color naming is complex, and by no means clarified as yet. In particular, it is not certain how far color categories depend on words. The "language" of honey bees, the best investigated social organism except man, does not contain any "color words." A scouting bee can communicate to her fellow foragers the abundance, odor, distance, and direction of a flower crop, but not its color (Ribbands, 1953). Nevertheless, the individual foragers' color perceptions can be shown in training experiments to be organized according to a few well-defined categories, which can be defined by the use of spectral light (Daumer, 1956) as well as on a more sophisticated relational scale (Kalmus, 1960).

The unconscious masking of sensory defects, aside from its specific dangers, is probably socially beneficial. But well-meant deliberate attempts to treat sense-deprived people as one would treat normal people sometimes assume rather absurd forms. During the last decade, for instance, I have read several reports in the popular press of pupils of the schools for the deaf performing in concerts and enjoying it. Not much was said about the listeners except that some parents of the children were gratified by the performance. It is, however, a fair bet that the playing of string and wind instruments and the choral production of such children can never be brought to a reasonable degree of perfection, and attempts of this kind are rather pointless. However, keyboard instruments and percussion may perhaps provide deaf children with some genuine satisfaction.

Minor sensory deprivations do not entirely destroy the appreciation of art. There is no doubt that some quite tune-deaf people find considerable enjoyment in listening to symphonic music and that color-blind people can greatly appreciate paintings. In neither instance does the deprivation destroy all the characteristics of the artistic structure contemplated. Nevertheless, as in either situation a full appreciation is eliminated by the relevant defect, one should as a rule counsel color-blind children not to paint but rather to play or sing and tune-deaf children to do the opposite.

To what extent sense deprivation can curb creative writing is not well known. Blind people have certainly written a great deal of prose and verse (Homer, Sappho, Milton), and so have people who have lost their hearing late in life. I do not know of any famous writer who was born deaf, but this may just be my ignorance. The American satirist James Thurber went on writing and even drawing when almost blind.

The ordinary writing of people born deaf or deafened early in life is frequently done in a sort of telegram style or special jargon, sometimes showing a loss of inflections and other grammatical deficiencies. Some of these features can probably be traced to insufficient schooling, but others may be more specifically attributed to the methods of communication that the teachers of the deaf and the deaf use among themselves. In particular these may be the use of pictures almost like ideograms in teaching and the sign language. A certain amount of knowledge on this topic exists in the schools and other organizations for the deaf, but a systematic investigation might be most interesting and might also contribute to an assessment of the relative roles which the spoken and the written word play in the development of an individual's language.

Any deficiency in "language" is bound to have other marked behavioral effects of great complexity. These could be expounded in a voluminous book in which the behavior of people suffering from every kind of sense deprivation would be separately analyzed by way of a monographic description, but this short paper cannot do so. Some general aspects deserving further consideration are: (1) the comparative sociology of the deaf and blind, especially their differences in mating habits; (2) differences from normal people and between themselves of character and emotional states of the variously sense deprived; and (3) the dependence of the adaptations and compensations of the sensory deprived on social conditions.

BIBLIOGRAPHY

BARTH, R.
1952. Estudos sobre os orgaos odoriferos de algunias Hesperidos Brasilians. *Mem. Inst. Oswaldo Cruz,* 50:423–556.

BUTLER, C. G.
1964. Pheromones in sexual processes in insects. *Roy. Ent. Soc. Symp.* 2:66–77.

CHAPANIS, A.
1951. Color blindness. *Sci. Amer.,* 184:48–53.

CRAWFORD, J.
1952. Living without a balancing mechanism. In M. Pinner and B. Miller (Eds.), *When doctors are patients.* New York: W. W. Norton.

DAUMER, K.
1956. Reizmetrische Untersuchungen des Farbensehens der Bienen. *Z. Vergl. Physiol.,* 38:418–451.

FISCHER, R., F. GRIFFIN, S. ENGLAND, and S. M. GAREN
1961. Taste thresholds and food dislikes. *Nature,* 191:1328.

GALTON, F.
1907. *Inquiries into human faculty and its development.* London: Dent.

HALLGREN, B.
1959. Retinitis pigmentosa continued with congenital deafness; with vestibulocerebellar ataxia and mental defect in a proportion of cases. *Acta psychiatr. et neurol. Scand. Suppl.* 133, Vol. 34.

HANHART, E., and R. LUCHSINGER
1951. Über die Bedeutung der erbmässigen Gleich-bzw. Ungleichheit (Homo- und Heterogenie) . . . *Schweiz. Med. Wochenschr.,* 81:726–731.

HOPKINS, L. A., and R. P. GUILDER

1949. Clarke School Studies concerning the heredity of deafness. Northampton, Mass.

JEFFERSON, J. P.

1958. A white-eyed mutant form of the American cockroach *Periplaneta americans*. *Nature*, 182:892.

KALMUS, H.

1943. The optomotor responses of some eye mutants of *Drosophila*. *J. Genet.*, 45:206–213.

1948. Tune deafness and its inheritance. *Proc. 8th Int. Congr. Genet.*, Stockholm.

1960. The sensory equipment of social hymenoptera. *Symp. Zool. Soc. London*, No. 3.

1961. The attenuation of optomotor responses in white-eyed mutants of *musca domestica* and of *Coelopa frigida*. *Vision Research*, 1:192–197.

1964. Genetical taste polymorphism and thyroid disease. *Trans 2d Int. Conf. Human Genet.*, Rome.

1965. Diagnosis and genetics of defective colour vision. Oxford: Pergamon Press.

KELLER, HELEN A.

1927. Sourde, muette, aveugle. Paris: Payot.

LAKOWSKI, R.

1962. Testing of colour vision in prospective printer's apprentices, and the problems this presents in selection. *6em Journees Int. de Couleur*, Eviane.

PHILIP, U., J. H. RENDEL, H. SPURWAY, and J. B. S. HALDANE

1944. Genetics and karyology of *Drosophila subobscura*. *Nature*, 154:260–263.

PICKFORD, R. W.

1955. Weak and anomalous colour vision in industry and the need for adequate tests. *Occup. Psychol.*, 29:182–192.

POST, R. H.

1962. Population differences in red and green color vision deficiency: a review and a query on selection relaxation. *Eugen. Quar.* 9:131–146.

RAY, V. F.

1953. Human color perception and behavioral response. *Trans. N.Y. Acad. Sci.* ser. II, 16:98–104.

RIBBANDS, R. C.

1953. The inability of honey bees to communicate colour. *Brit. J. Animal Beh.*, 1:5–7.

RICHTER, M.

1953. Ergebnisse von Farbensehprufüngen in färbrischen Betrieben. *Die Farbe*, 2:175–192.

RIDDELL, W. J. B.

1940. A pedigree of hereditary sex-linked night blindness. *Ann. Eugen.*, 10:326–331.

TREVOR-ROPER, P. D.

1957. The influence of eye disease on pictorial art. *Proc. Roy. Soc. Med.*, 52:721–744.

VON UEXKÜLL, J.

1921. *Unwelt und Innenwelt der Tiere*. Berlin: Springer.

SELECTION AND GENETIC HETEROGENEITY

J. M. THODAY

IN RECENT YEARS students of natural and of artificial populations have demonstrated enormous genetic heterogeneity in such populations and have come to realize that there are many factors that will maintain such genetic heterogeneity. Earlier, though the classical mathematicians of the subject, Fisher, Haldane, and Wright, had discussed selection in clines, density-dependent factors, and the effects of gene migration between different adaptive peaks, more simpleminded biologists concluded from the mathematics that natural populations might be expected to be genetically rather uniform; such variety as they envisaged was regarded as dependent largely on recurrent mutation.

Experimental work on polymorphic populations and its extension to the study of continuous variation showed this view to be erroneous. Polymorphic populations containing two or more clear-cut phenotypes at first appeared to provide exceptions to a general rule. We now know they are but the striking examples of a general rule. Natural populations of outbreeding species, including those of man as studies on homografts have made especially clear (see Medawar, 1957), far from being genetically uniform, are genetically so heterogeneous that without danger of exaggeration we may regard each individual as genetically unique, with, of course, the notable exception of identical twins.

These discoveries have enormous implications for our understanding of man. As human individuals, we have to realize that we must each adjust to living with men different from ourselves, even within the family, and specialists, who study men from some particular point of view—the biochemist, the physiologist, the psychologist, and so forth—have to realize that in studying normal man, even in a closely confined population, they are studying a highly heterogeneous set of genotypes—even if it may *appear* a comparatively uniform set of phenotypes.

Critical demonstration of the types of genetical difference involved is, of course, much more practicable with experimental organisms. Before discussing man further, we shall therefore consider something of the causes and functions of this widespread genetic diversity and something of its genetic nature as revealed by studies of experimentally more amenable organisms.

Two factors help us to explain this diversity. First, studies of clearly polymorphic populations have shown us that natural selection pressures, earlier without good reason always supposed to be low, may be very high. For instance, Allison (1955) has estimated the relative viabilities of the sickle-cell heterozygote in man in malarial areas as 1.26 times that of the normal homozygote, a 26 per cent difference, and Dowdeswell, Ford, and McWhirter (1960) have estimated selective elimination as high as 60 per cent in *Maniola jurtina*, whereas up to 1955 population geneticists thought in terms of selection coefficients of the order of 1 per cent. Second, we now have more sophisticated ideas of the variety of forces—some selective—which may maintain heterogeneity and hence also of the functions of heterogeneity. These are as follows:

1. Long-term selection for adaptability (as opposed to adaptation) operating in a changing environment to preserve the outbreeding systems on which heterogeneity depends (see, for example, Darlington, 1939; Mather, 1943; Thoday, 1953, 1958).

2. Short-term selection for heterogeneity in a heterogeneous environment; that is, centrifugal (Simpson), disruptive (Mather), or diversifying (Dobzhansky) selection.

3. Heterozygous advantage.

4. Assortative or disassortative mating.

All these may promote and preserve heterogeneity, though the last two may often be consequences of the first two and hence secondary phenomena. Disruptive selection is the most recent of these to receive intensive investigation, and experiment (Thoday, 1958–1965) has shown it to have powerful effects. Our new knowledge of the strength of natural selection should make us the more ready to expect powerful effects in nature.

We speak of disruptive selection when selection picks out within a single population more than one class of individual, rather than favoring only one. (If it favors one end of a distribution curve, we speak of directional selection; if it favors the mean value of a character at the expense of deviants in both plus and minus directions, we speak of stabilizing selection.) The simplest form of disruptive selection is that which favors both extremes equally. Disruptive selection will occur whenever an interbreeding population is exposed to heterogeneous environmental conditions and will then influence any genetic loci that affect fitness in those conditions. The heterogeneity of the environment may be spatial or short-term temporal or both.

The experiments referred to have used *Drosophila* as material, the character being sternopleural chaeta number. They have shown unequivocally that disruptive selection can:

1. Increase phenotypic diversity by increasing genetic variance, and with it genetic flexibility or the genetic adaptability of populations (Thoday, 1959).

2. Establish clear-cut polymorphisms (Thoday and Boam, 1959; Thoday, 1960; Millicent and Thoday, 1961).

3. Establish and maintain distinct differences between parts of a population that exchange genes through hybridization at rates equivalent to and even higher than that which must occur with random mating (Thoday and Boam, 1959; Millicent and Thoday, 1961).

4. Split a population into two distinct phenotypically different populations that

are reproductively isolated from one another, that is to say, do not exchange genes (Gibson and Thoday, 1963, 1964; Thoday and Gibson, 1962).

The first of the effects of disruptive selection to be demonstrated concerned the diversity of a population in an experiment in which disruptive selection increased and stabilizing selection decreased the variety of individuals by comparison with the original wild stock from which the experiment was begun.

It is of some interest to consider what we know of the genetic means whereby this result was achieved (Gibson and Thoday, 1962a, 1962b; Wolstenholme and Thoday, 1963). Three classes of chromosome II were found in the disruptive selection line. They differed at two genetic loci. The alleles increasing chaeta number will be referred to as + alleles, those decreasing chaeta number relatively to + as − alleles. The three classes of chromosome were + −, − +, and − −. − −, heterozygous with either of the others, was responsible for the low chaeta number flies established in the population by disruptive selection. The population was in fact polymorphic at these loci. Now when we had established the existence of these loci, we were able to look for them in the wild stock with which the experiment began. Forty-eight chromosomes proved to comprise 46 + − and 2 − +, the + −/− + heterozygotes being capable of generating − − by recombination, the loci being 20 map units apart.

We have since (Gibson, unpublished; Thoday, 1965) tested wild stocks from the United States, Spain, Italy, Israel, and Japan (the original wild stock having been captured in Britain) and found both + − and − + in all but those obtained from Israel. Here is critical proof in terms of specific genes affecting a classic "polygenic" character of the kind of genetic diversity we know to be widespread as a result of the work of many students of natural and artificial populations, notably Dobzhansky and Mather. It is diversity within the *normal* range. With respect to these loci, there are two kinds of normal homozygote and one kind of normal heterozygote. The two kinds of homozygote are normal for different genetic reasons. The heterozygote is normal, but has the potentiality of producing extremes by recombination. Disruptive selection exploited this potential diversity by establishing the − − chromosomes as normal in the new, experimental situation.

In a further experiment (Thoday, 1960), disruptive selection established and maintained a rather clear-cut polymorphism involving a polygenic character. The selection was again for chaeta number, but the lines used had marker genes associated with the chaeta number genes. Genetic analysis of the chromosomes involved has shown us that the polymorphism was as follows:

High selected flies		Low selected flies
e + + +		e se − − − cp
———————————	X	———————————
e se − − − cp		e se − − − cp

e is a gene enhancing the effectiveness on chaeta number of the + + +/− − − vs. − − −/− − − genetic difference. se and cp are "marker" genes, with no effect on chaeta number, that affect eye color and wing shape respectively. These were maintained polymorphic solely due to their linkage to the chaeta number loci that were themselves directly affected by the selection.

This illustrates not only how a complex polymorphism may be maintained by heterogeneous selection pressures, but that irrelevant genes giving a more conspicuous polymorphism may be carried along through their linkage to the genes more directly affected by selection. We must not therefore expect that all observed polymorphisms will be the direct effect of selection. Complex interlocking of different genetic systems may often be responsible. Behavioral polymorphisms might be maintained by selection for morphological or physiological polymorphism, or vice versa.

Some further experiments concern a different question, the establishment of differentiated classes by disruptive selection. We have shown that disruptive selection can readily bring about differentiation of genetic classes within a population (Thoday and Boam, 1959; Millicent and Thoday, 1961). In two experiments, this differentiation extended to establishment of isolation. This occurred in a disruptive selection experiment in which the differing classes of fly selected were given choice of mates (Thoday and Gibson, 1962; see also Thoday, 1965). Very rapidly (by twelve generations in one line and seven in another) assortative effective mating was established such that most of the progeny were formed by mating high chaeta number flies or between low chaeta number flies, and very few by mating of high with low or low with high. It is not certain yet that this was a behavior phenomenon, though there is evidence that differential mating choice must have been involved as well as differential mating success. Assortative mating is well known in man and the astonishing rapidity with which these experiments produced these results seems relevant.

All populations are exposed to heterogeneous selection pressures and hence to disruptive selection. It is easy to think of examples of this for man. Differential housing conditions and other economic factors induce differential selection for any genes conducive to resistance to tuberculosis. Housing conditions and economic factors change rapidly with time, as does the genotype of the tuberculosis pathogen and the mode of medical treatment. We must expect human populations in appropriate areas to be polymorphic for many loci affecting resistance to tuberculosis, for any pleiotropic effects of these loci, and for some closely linked loci affecting other characters. Some of these pleiotropic effects and linked loci must affect behavior. We face an exceedingly complex situation. Again, there is strong evidence (Spickett, 1964) that both the incidence and form of leprosy are greatly affected by genetic factors. There is no need to stress the behavioral consequences of leprosy, both for those suffering it and for the other members of the society to which they belong. It is in fact likely that most genetic variables have some behavioral consequences. In mice in my laboratory at present, Spickett and others are demonstrating great genetic variance affecting adrenocortical steroid metabolism and associated physiological and histological characters. The influence of these substances on behavior is beyond question. Tanner (1965) has discussed behavioral consequences of many physical variables that vary genetically in man.

We must therefore expect that most variables will have genetic components and that most variables will have behavioral consequences. I want to consider what this means from the point of view of behavior studies and of our general attitude.

Behavior characters are complex characters. Understanding of them will require that they be analyzed. Satisfactory analysis will never be achieved by those who take, as have so many geneticists and psychologists in the past, extreme geneticist or environmentalist views. Complex characters will be "polygenic" characters. No such character in any natural population of any outbreeding species has been critically studied, whether it concerns behavior or not, without evidence being found for *both* genetically and environmentally caused variation. We must study both the genetic variation and the environmental variation in the population under study.

Now it is not unusual, when the variation of a character is in part heritable but proves to be "polygenically" controlled, for the investigator to give up the genetic side of the study as unprofitable.

Recently, however, we have shown that this need not be so. On the contrary, further study of the genetics may greatly aid understanding of the character from a physiological point of view. At least in as good an experimental organism as *Drosophila* is for the purpose, it is possible to locate on linkage maps some of the relevant genes (Thoday, 1961). Having done so, it is then possible to use the information for new analysis of the complex character.

Thus Thoday, Gibson, and Spickett (1964) showed that a difference in chaeta number between two stocks—one with 20, the other with 35 chaetae/fly—was largely due to three genes. One gene was then shown to affect chaeta number everywhere by increasing cell number. Another had a local effect by delaying initiation of chaeta development at one particular place with the result that several chaetae initiated later. The third gene had less understood effects, but these were confined to a different localized region. Thus the genetic analysis of chaeta number made possible the analysis of chaeta *pattern* into certain unit components which had not been possible by studying pattern and number themselves.

There is no reason why similar studies should not be used to aid analysis of complex behavior patterns in *Drosophila* and the mouse. The latter studies might suggest new ways of analyzing behavior patterns in man and thus, by providing clues to unit components of human behavior patterns, aid the search for their genetic control. We must, furthermore, expect to find genetic variation underlying apparent similarity in behavior pattern.

I have referred above to the type of genetic system that may be involved in such a situation in discussing the three normal chaeta number genotypes

$+ - / + -$, $+ - / - +$, and $- + / - +$ of *Drosophila* we have discovered. More revealing is an analysis of some other lines made in our laboratory by Spickett (1963). He analyzed the polygenic systems distinguishing two lines of flies one with very high, the other with normal, chaeta number. Both lines had the same average size or weight of fly. He found that one of the high chaeta-number polygenes was associated with high cell number, giving a large fly. But the effects of this were exactly compensated for by another gene, which had no effects on chaeta number but made small cells. He thus had two kinds of normal-size fly. One $+ - / + -$ had relatively few large cells; the other $- + / - +$ had relatively many small cells. Both are balanced, or compensated, genotypes.

We have evidence of a similar distinction between different strains of mice in which more active synthesis of adrenocortical steroids is compensated for by more active breakdown of the steroids and vice versa. Such compensated systems may give individuals which, with our crude techniques of character analysis, seem the same, but which are really very different in the architecture of their morphological, physiological, and behavioral patterns. We have good reason to believe that such balanced polygenic systems are very common, if not ubiquitous.

The genetical analyses that made these exercises possible used marker genes and determined the linkage of chaeta number or fly size to these. But polygenic characters are usually studied biometrically. We now have plans to design new combinations of these two techniques to make the analyses of polygenic variation easier. Once this has been done, it may prove possible to modify the biometrical techniques so that populations may be studied in order to reveal the linkage of component genes of polygenic systems to the markers that naturally segregate in the populations. It may then be possible to use such techniques to throw light on the genes mediating variation in complex characters in man, such as IQ or behavior pattern, and to discover something of the unit components of these characters.

Meanwhile we are forced to accept that the ubiquitous genetic diversity we know to exist will affect psychological as well as physical characteristics in man, despite the difficulty we have in sorting out biological and social inheritance components for such characters in human populations. We must, I submit, face the fact that each of us is in some way, probably in many ways, genetically—hence biochemically, physiologically, and psychologically—different from all others.

It is difficult for us to adjust to this concept. One can readily see that even scientists who should find such adjustment relatively easy do not. The literature of psychology, social psychology, sociology, and education is riddled with one-sidedness in this matter, and those who can dispassionately avoid the extreme geneticist or environmentalist views are rare. In lay discussions of selection for education, of race, and of class distinction, the antithesis heredity *or* environment is general.

The human situation is exceedingly complex, and the problems of distinguishing between the effects of biological heredity and the effects of social heredity are so difficult that it is hardly surprising that controversy arises whenever the obvious view that both are always concerned is stressed. And the concept that the question "Is this variety inherited or environmental?" has no meaning is accepted by so few that we cannot expect the likely interactions to be appreciated.

Yet we must expect that every complex "character" will show genetic as well as environmental variance and that the genetic variance will be subject to change as a result of selective and other forces that themselves change with social change. We are not here talking about major classifiable variation such as is mediated by major "Mendelian" mutant genes, though these also are of great importance. We are rather concerned with variation within the *normal* range. Here we must expect many genetic consequences of social forces and social consequences of a population's genetic structure and of changes in that structure.

IQ, for example, is a complex character, which provides an example more profitably discussed than most because it is a relatively precisely measurable quantity, correlated with many factors of performance in life. That there is a heritable component of IQ there is no doubt. The usual estimates are of the order of 0.5,[1] that is to say, about half the variance of IQ is the consequence of genetic variety. Those with higher IQ's tend, especially in countries with free education to the university level, to rise in the social and economic scale. With a heritability of 0.5, the progeny of those parents with higher than average IQ will be less genetically well endowed in this respect than their parents. In fact, we would expect their genetic endowment on the average to be about halfway between their parents' mean and the population mean. This will be partially compensated for at the phenotypic level by the higher IQ parents' handing on through social heredity a better environment, but the genetic component of IQ would be unaffected by this. The inevitable consequence of such a situation must be some rise of higher IQ genotypes from the economically lower classes in each generation, compensated for by fall of lower IQ genotypes from the higher classes, with a lag in the number of generations which high IQ families take to fall in the social scale owing to the social heredity component. This process will give rise to a measure of assortative mating for IQ that itself will contribute to the maintenance of IQ genotypic variance.

More interesting and more complex are the questions that arise if the character, in this case IQ, is correlated with fertility. There has been a great deal of discussion arising from the demonstrations that IQ is negatively correlated with family size, and the consequent prediction that mean IQ must decline over the genera-

---

1. It cannot be too often stressed that the heritability of a character is not a measure applicable in general but only to the particular population from which the measure was obtained, in the variety of environmental circumstances in which the individuals of that population grew up. An increase of the variety of environmental conditions will decrease heritability just as will a decrease of genetic variety. A more uniform environment will increase heritability. We must never expect a single value for the heritability of a character in a species.

tions. The investigations of Godfrey Thompson threw doubt on this prediction, for they failed to detect the predicted decline. More recently it has been shown that the situation is more complex than was hitherto thought. The negative correlation within a generation is real, but the method of investigation necessarily ignores zero size families.

When these are taken into account, which involves investigating two generations at least, it appears that the negative correlation is about compensated for by a positive correlation between IQ and the probability of having any family at all (Willoughby and Coogan, 1940; Higgins, Reed, and Reed 1962; Gibson, unpublished; see Newcombe, 1963, for references and for important discussions of the need for studies involving more than one generation). This leaves us with a new negative correlation, not between IQ and mean family size, but between IQ and variance of family size, and means that the genetic situation is rather labile. The low probability, at the low end of the IQ range, of marrying or having any children may depend on many factors. Perhaps high mortality and lesser physical attractiveness to mates may be involved: so will, in all probability, less skillful use of cosmetics, less care of teeth, and so on. All these are open to improvement by improved medical care, education, and improving general economic conditions. If these were effectively improved, the variance of family size would go down, the mean family size at the low end of the IQ scale would go up, and once again we would have to predict a decline of mean IQ of the population as a whole.

I discuss this example, not because I want to make this prediction, but because it illustrates rather well that, once we accept that variables will have *both* genetic *and* environmental components, we must expect not only behavioral consequences of genetic variation in man, but also genetic consequences of behavioral variation in man.

All our evidence from man and other outbreeding organisms, and even from so-called inbreeders, tells us that each population contains immense genetic variety, and that genetic variance, environmental variance, and genotype-environment interaction will each be concerned with every complex variable. Our problem is to discover what proportion of the variance is genetic or environmental in any particular case. We must then envisage that the genetic variance will have environmental (including social) consequences, and the environmental variance will have genetic consequences. In man genotype may influence environment as well as vice versa. Highly polymorphic populations need polymorphic environments. In an organism with control of the environment, polymorphic populations will and should make polymorphic environments. Social misfits will be those who have not had appropriate environments, often because of men's propensity for trying to make other, different men like themselves. We should become aware that people can only be alike if each has an appropriate unique environment.

Our demonstrations of the efficacy of disruptive selection—not only in producing genetic variety and polymorphism, but also in establishing genetic differences between parts of a population despite extensive gene exchange—must reinforce our expectation that groups of all kinds such as are formed in human societies may differ in gene frequencies, and that there will be continual interplay between social structures with genetic consequences and genetic differentiation with social consequences. It is our task, not to question this, but to find means of understanding this interplay in useful detail.

BIBLIOGRAPHY

ALLISON, A. C.
1955. Aspects of polymorphism in man. *Cold Spr. Harb. Symp. quant. Biol. XX*, 239–255.

DARLINGTON, C. D.
1939. *The evolution of genetic systems*. Cambridge: Cambridge University Press.

DOWDESWELL, W. H., E. B. FORD, and K. G. McWHIRTER
1960. Further studies on the evolution of *Maniola jurtina* in the Isles of Scilly. *Heredity*, 14:333–364.

GIBSON, J. B., and J. M. THODAY
1962a. Effects of disruptive selection. VI. A second chromosome polymorphism. *Heredity*, 17:1–26.

1962b. An apparent 20 map-unit position effect. *Nature*, 196:661–662.

1963. Effects of disruptive selection. VIII. Imposed quasi-random mating. *Heredity*, 18:513–524.

1964. Effects of disruptive selection. IX. Low selection intensity. *Heredity*, 19: 125–130.

HIGGINS, J. V., E. W. REED, and S. C. REED
1962. Intelligence and family size: a paradox resolved. *Eugen. Quart.* 9:84–90.

MATHER, K.
1943. Polygenic inheritance and natural selection. *Biol. Rev.*, 18:32–64.

MEDAWAR, P. B.
1957. *The uniqueness of the individual*. London: Methuen.

MILLICENT, E., and J. M. THODAY
1961. Effects of disruptive selection. IV. Gene-flow and divergence. *Heredity*, 16:199–217.

NEWCOMBE, H. B.
1963. Intelligence and genetic trends. *Science*, 141:1104–1109.

SPICKETT, S. G.
1963. Genetic and developmental studies of a quantitative character. *Nature*, 199: 870–873.

1964. Genetic mechanisms in leprosy. In R. G. Cochrane and T. F. Davey (Eds.), *Leprosy in theory and practice*. Bristol: Wright.

TANNER, J. M.
1964. Human growth and constitution. In G. A. Harrison, J. S. Weiner, J. M. Tanner and N. A. Barnicot, *Human biology*. Part IV. Oxford: Clarendon Press.

THODAY, J. M.
1953. Components of fitness. *Symp. Soc. exp. Biol.*, 7:96–113.

1958. Natural selection and biological progress. In S. A. Barnett (Ed.), *A century of Darwin*. London: Heinemann.

1959. Effects of disruptive selection. I. Genetic flexibility. *Heredity*, 13:187–203.

1960. Effects of disruptive selection. III. Coupling and repulsion. *Heredity*, 14:35–49.

1961. Location of polygenes. *Nature*, 191:368–370.

1965. Effects of selection for genetic diversity. *Proc. XI Int. Congr. Genet.* The Hague, Vol. 3, 533–540.

THODAY, J. M., and T. B. BOAM
1959. Effects of disruptive selection. II. Polymorphism and divergence without isolation. *Heredity*, 13:205–218.

THODAY, J. M., and J. B. GIBSON
1962. Isolation by disruptive selection. *Nature*, 193:1164–1166.

THODAY, J. M., J. B. GIBSON, and S. G. SPICKETT
1964. Regular responses to selection. II. Recombination and accelerated response. *Genet. Res.*, 5:1–19.

WILLOUGHBY, R. R., and M. COOGAN
1940. The correlation between intelligence and fertility. *Hum. Biol.*, 12:114–119.

WOLSTENHOLME, D. R., and J. M. THODAY
1963. Effects of disruptive selection. VII. A third chromosome polymorphism. *Heredity*, 18:413–431.

HEREDITARY FACTORS IN PSYCHOLOGICAL VARIABLES IN MAN, WITH A SPECIAL EMPHASIS ON COGNITION

STEVEN G. VANDENBERG

IN TRYING TO SUMMARIZE a lot of information about hereditary factors in human psychological differences, one is overwhelmed by the mass of details that demand some organization. In their admirable book on *Behavior Genetics* (1960), Fuller and Thompson set a difficult standard to follow. This paper will concentrate on material not available at that time, most of it from twin studies. No attempt will be made to present details on the many psychological tests employed in the studies reviewed here, because to do so would make this paper far too long. The interested reader is referred to the references given for these studies.

One way to organize the material might be to use the extent to which a trait can be modified by training or other environmental factors as an axis, starting from highly plastic traits through moderately stable traits to traits that appear to be fixed for life, and to see if the evidence for hereditary factors increases as one would expect. Unfortunately, I do not know enough about the modifiability of many of the variables to use this approach. Instead, I followed this scheme: findings are ordered along a dimension that might be thought of as running from purely structural anatomical variables, through physiological variables in which some central factor plays a role, to what one might call purely psychological variables, such as attitudes and value judgments.

If we think of this continuum as one that shows an increasing involvement of the personality as opposed to the physical equipment, we might expect a progressive decline in the importance of the hereditary component. This is, to some extent, what the results show. Exceptions to this rule may be made more understandable if we rephrase the sentence somewhat and think of this continuum as concerned with an increasing involvement of the central nervous system as against more peripheral structures. Structural and more or less permanent functional differences may exist between individual nervous systems, which are themselves genetic in origin and which may affect certain aspects of personality more than other aspects; this would explain apparent departures from the expected progressive decline in importance of hereditary factors as we move "in" and "up."

To avoid awkward circumlocutions I will frequently use the term "heritability" as shorthand for the statement that a trait is under some degree of hereditary control, even though no precise measure of heritability exists for man.

The conclusion that a trait shows heritability or some amount of heritability is based on statistical evidence, so that we encounter the usual questions of whether the sample is representative of the population, whether the mathematical model bears an adequate relation to physical reality, how efficient the statistic is, and so forth. I will not discuss these topics, mostly because I am not competent to do so. I will merely mention here the most frequently used techniques for deciding whether a trait has an important hereditary component. Holzinger devised an heritability index, h^2, of the proportion of the variance due to heredity based either on subtracting correlations from one another or in another formula subtracting variances from one another. The index has been criticised as statistically and genetically not quite correct. As serious perhaps is the fact (for which the author cannot be blamed) that the index is frequently misinterpreted, so that results of studies that used this index are frequently misquoted. A more modern and probably less easily misquoted method consists of a comparison by Fisher's F-test of the within-pair variances of identical and fraternal twins. This method has the added attractiveness that one can further partition the variances by considering between occasions and between tests variances. A disadvantage may lie in its lack of intuitive clarity to those not used to statistical thinking. Nor does it give a measure of the relative importance of heredity or of environment for that trait. Perhaps this is just as well, because this relation would differ from sample to sample, and because such a measure may have no meaning unless it is stated in statistical terms, because the development of all human traits requires in a very real sense a normal environment and a normal complement of genes within a very narrow range of variation. The studies will be reviewed in this order: (1) anatomical and anthropometric, (2) sensory, (3) perceptual, (4) motor skills, (5) cognition, and (6) personality—temperament, ANS reactions, attitudes and interests, and values.

ANATOMICAL AND ANTHROPOMETRIC

I include this category mainly for the sake of completeness, and will only mention some traits in which I have been interested. There have been extensive anthropometric studies of twins by Dahlberg (1926), von Verschuer (1954), Clark (1956), Vogel and Wendt (1956), Osborne and DeGeorge (1959), Vandenberg and Strandskov (1964), and Conterio and Chiarelli (1962). They show, in general, very good agreement (Vandenberg, 1962b), so that it may be better to stop doing studies of this kind unless refinements of techniques or new ways of analyzing the data are forthcoming. Multivariate analyses might eliminate the need to take the many measures employed in these studies.

Lundström (1955) has estimated the heritability of facial angles; studies of dental dimensions in twins were reported by Lundström (1948), Hunter (1959), Horowitz, Osborne and DeGeorge (1958), and Vogel and Reiser (1960), while similar studies are in progress in the University of Oregon Dental School (Savara, 1963, 1965) and in Richmond, Virginia (Hughes, personal communication). There appears to be a tendency for heritability estimates to be higher for dimensions of teeth which erupt earlier. Is it meaningful to think of these teeth as less influenced by the postnatal environmental factors?

Sarah Holt (1961), Pons (1958), and others continue to work on quantitative methods for the genetics of fingerprints. Spuhler (1951) reported on genetic components in the patterning of the superficial veins on the chest, in variation in the peroneous tertius muscle that everts and dorsoflexes the foot, and in the number of vallate papillae. Lysell (1955) studied, in twins, the length and width of the papillae and the direction of ridges on the palate and obtained high heritability estimates for the former, but not for the latter. Characteristics of the iris were studied in a small number of twins by Ritter (1958), who obtained strong evidence for hereditary control. Many hereditary abnormalities are known in the structure of the eye, most of which affect its functioning. Nakajima (personal communication) measured ten dimensions of the eye interiors.

HEREDITARY FACTORS IN SENSORY VARIABLES

Hereditary factors are known in color vision, visual acuity, audition, perception of certain tastes, perhaps smell, and pain perception. In fact, logically one would expect them in all modalities.

Of the sensory capacities, let us first consider vision. Anomalous color vision has been known for a little over a century, which makes us wonder if other sensory anomalies are being overlooked. Several forms are now known to be controlled by genes located on the nonhomologous part of the X chromosome. Most of the studies have failed to distinguish between the various types of anomalies such as the protanopes, protanomalous, deuteranopes, and deuteranomalous because these defects are relatively infrequent in men, that is, 5 per cent for the last mentioned and 1 per cent each for the others. Because two mutant genes are necessary in women, the incidence is far lower. The most reasonable place of action of the genes would be on the retinal biochemistry.

Color anomalies are considered all-or-none phenomena. Pickford (1951) proposed that graduated individual differences in color vision exist. If these turn out not to be due to inaccuracy, indecisiveness, and other personality factors, we will have to assume either that there exist other components in the total system for color vision than the retinal ones controlled by the sex-linked genes, such as perhaps some factors controlling the efficiency of the central nervous system integration somewhere along the optic pathway, or that the impairment of the retinal biochemistry can be partial and the anomalies are only on the extreme of a curve. Future studies of linkage with primaquine sensitivity and Christmas disease would be enhanced if more cases of the less frequent types could be found, perhaps by some mass screening technique. We may expect to see even more co-authors for such perfected studies than the recent papers in this area could boast of. Post (1962a) recently reviewed color anomaly rates in different parts of the world and reviewed the attractive hypothesis that changing social conditions allowed a relaxation of selection pressures. In a second article, Post (1962b) reviewed differences in vision acuity with the same hypothesis in mind. He concluded that the evidence is even more tenuous here but does support the idea that the frequency of vision defects in a population is proportional to the duration since its emergence from a "primitive" economy. Details of his work have been criticized, but this is the kind of theorizing that we need, if we are to trace the total path between gene and behavior.

The nature of the proposed selective advantage of myopia is an interesting question. Could it be an ability to do close work under somewhat poorer light than normals can accept? This could have been a real advantage through the ages, before the advent of electric lights.

I will deal even more briefly with the other modalities. Recent papers by Vandenberg (1962a) and by Diane Sank (1963) gave further evidence for genetic factors in audition respectively in normal and in deaf twins, which could be added to the summary in Fuller and Thompson (1960).

The possibility cannot be ruled out that the results of the normal twin studies do not reflect hereditary determination of normal individual differences but are, instead, caused by unrecognized beginning deafness in one member of some discordant twin pairs.

While PTC tasting is a well-studied genetic variable, diphenyl-guanidine (Snyder and Davidson, 1937) and brucine (Barrows, 1945) are less well studied, and perhaps because of technical difficulties there is little or no work going on toward discovering bimodal distributions for other substances. Lists of dislikes for food have been used in psychiatric studies; possibly this technique could be modified for preliminary studies of the genetics of human taste.

Practically nothing is known about hereditary factors in smell. Now that the possibility has been raised that molecular structure is related to the nature of the smell, one obstacle to more systematic work may have been removed.

Clinical reports of individuals without the sense of pain have been too rare for genetic analysis. Perhaps astronaut selection research has turned up some indication of differences in the efficiency of the inner ear or of kinesthetic sensation. The ease with which persons become seasick, carsick, and so forth, might also be explored to see if there is a genetic basis.

PERCEPTION

Fuller and Thompson grouped sensory and perceptual variables together and opposed them to response variables. This grouping probably reflected their feeling that sensory and perceptual performances are less affected by central conditions. Actually, there is not a really strong distinction. Perceptual variables may not only differ from sensory ones because the stimulus is more complex, but also perhaps because they are, in large part, under the control of mechanisms "located" at a slightly higher level in the nervous system judging by animal extirpation experiments, so that the organization within and between areas of the brain may be involved. In that case, a certain response-like feedback system may be involved. That effective organization of brain areas may, in part, be acquired on the basis of learning by the individual seems indicated by the findings that some animals reared in the dark lack functional vision.

Fuller and Thompson have reviewed studies on the apparent size of after images; the Müller-Lyer illusion in which the two figures

$$\longleftrightarrow \quad \text{and} \quad \rangle\!\!\longrightarrow\!\!\langle$$

are to be matched in length (von Bracken, 1939; Smith, 1949, 1953); the frequency of critical flicker fusion; the autokinetic response (Eysenck and Prell, 1951); and orientation in space (Malan, 1940). To this can be added evidence from a number of paper-and-pencil tests used in the Thurstone-Strandskov study (1955) and in the Michigan study. Some of these tests are commonly regarded as measures of intelligence but require, to a varying degree, the perception of patterns or of similarities in designs.

TABLE 1
F-RATIOS OF DZ AND MZ WITHIN-PAIR VARIANCES ON PERCEPTUAL TESTS.

	Thurstone et al. (1955)			Vandenberg (1962a)	
	53 DZ	43 MZ		37 DZ	45 MZ
1. Kohs' Blocks (Arthur scoring method)	4.57				2.53
2. Gottschaldt Figures	2.05				3.05
3. Concealed Figures (Thurstone)	3.05				
4. McGill Closure Test (Mooney)	2.31				
5. Hidden Pictures	2.27				
6. Hidden 4-Letter Words (Thurstone)					2.10
7. Street Gestalt Completion Test	2.05		Form A		1.63
			Form B		2.33
8. Mutilated Words (Thurstone)	1.69		Form A		1.37
			Form B		2.00
9. Incomplete Words (Thurstone)	1.81				
10. Dot counting	1.93				
11. Identical Forms (Thurstone)					1.60
12. Identical Numbers (Thurstone)					1.52
13. Backward Writing (Thurstone)	1.49				
14. Copying Designs (Thurstone)					1.48
15. P test	1.44				
16. Dot patterns	1.17				
17. Cross-out	1.05				

Table 1 shows the results for these tests in order of descending significance of the F-ratio. The few replications show a fair degree of consistency. The highest values appear to be for tests calling for a more complex perceptual process involving a considerable exercise of deliberate judgment, especially in the Kohs' block designs test which is, after all, one of the subtests in Wechsler's Intelligence Test. It will be interesting to see to what extent these tests measure the same thing.

An especially stringent test will be one based on the correlations of twin differences on these tests. Some of these correlations may be found in Appendix 1.

MOTOR PERFORMANCE

Additional material might be added to Fuller and Thompson's summing up of the evidence for a considerable hereditary component in the area of motor skills.

Tapping speed shows an hereditary component in a number of studies (Newman, Freeman, and Holzinger, 1937; Eysenck and Prell, 1951; Thurstone *et al.*, 1955).

Table 2 compares the results for the right hand and the left hand on a number of other motor skill tasks. The first thing to be noted is the high incidence of significant evidence for heritability. It appears as though almost any task involving motor skills is affected by hereditary factors. We know from studies by Fleishman (1960) and others that motor skill tests do not measure a general factor: the correlations between them are generally low and a rather large number of independent factors have been isolated. We cannot say at this point to what extent these are due to independent hereditary components.

The next comment on Table 2 concerns the fact that there appear to be some inconsistencies when comparing the results for the two hands. Whether these reflect lack of precision of the measures, or unreliability of twin differences or real discrepancies in heritability can only be decided by experiments especially designed to study such factors.

TABLE 2

F-RATIOS BETWEEN DZ AND MZ WITHIN-PAIR VARIANCES ON
ELEVEN MOTOR SKILL TESTS FOR THE RIGHT HAND AND FOR THE LEFT HAND.

Thurstone et al. (1955)	Right hand	Left hand
1. Dynamometer	3.68**	2.17
2. Minnesota Placing	2.25**	2.10**
3. Tapping	1.34	3.86**
4. Telegraph key	2.59**	.98
5. Two-hand coordination	1.99**	1.53**
Vandenberg (1962a)		
1. Tweezer dexterity	3.40**	2.73**
2. Card sorting	2.57**	3.42**
3. Mirror drawing	3.38**	1.31
4. Santa Ana dexterity	2.41**	1.05
5. Rotary Pursuit	2.08**	1.46
6. Hand steadiness	1.59	1.21

** indicates significance beyond the .01 level of probability.

"INTELLIGENCE"

I shall not review the kind of evidence that indicates that intelligence is in part controlled by hereditary factors nor the controversies on how to interpret the evidence. Nor shall I repeatedly qualify my remarks by reminders that I, too, believe that the environment plays an extremely important role in the development of intelligence.

What I should like to do instead is to discuss briefly some newer (and some not so new) approaches to the measurement of intelligence and to see whether they may be relevant to the study of human genetics.

When Alfred Binet developed the first effective intelligence test, he had no theory of intelligence to guide his selection of items. It is ironic that fifty years later we still have no one satisfactory theory that is generally accepted. No doubt this is partly due to the fact that this is really a very tough problem, but it is perhaps as much due to the fact that the very success of the Binet test, and its American adaptations, for many years made intelligence a less highly regarded subject matter for research than, say, learning theory. I mention learning theory deliberately, because a complete theory of intelligence will some day have to include learning, memory, reasoning and problem solving, language and speech, and so forth in an interrelated system of concepts. At the present time these are almost completely separate areas which are taught separately and in which research is done without much interaction with one another, let alone with research on intelligence. Two recent books edited by Harper, Anderson, Christensen, and Hunka (1964) and by Constance Scheerer (1964) bring together a variety of viewpoints that will need to be integrated into a comprehensive theory of cognition. Up to now, such theories have neglected differences between subjects in their search for general laws (Hirsch, 1963). But the essence of measuring intelligence is recognition of individual differences. A satisfactory cognitive theory should include consideration of individual differences as an integral part of the conception, not added as an appendix. Until recently, theories about intelligence have been dominated by one single question: whether there is such a thing as a general level of intelligence that can be expressed in one number, such as the IQ, or whether there exist a number of independent abilities so that a person could do well on some ability tests and poorly on others.

For diagnosis in clinic, hospital, and school and for military, industrial, and business selection and placement, a single number was so easy to use, while still providing information not available elsewhere, that it was widely accepted. This is not to say that there were not early misgivings about the overly simple nature of such measures. Criticism has taken many forms but perhaps one can distinguish two main trends. The first might be called *clinical* or *dynamic*. This type of criticism is concerned with the fact that a person may not always do equally well even on the same test. This variability is seen not so much as a lack of reliability on the part of the test but more basically as an attribute of the person. An extreme form of this type of thinking is encountered when the performance on part of a test is used to estimate the potential level of the subject, while the variability of performances on other parts around this level, called "scatter," is taken as a measure of deterioration or regression. Because knowledge of words is least affected by changes in mood, test-taking attitudes, and clarity of mental processes, vocabulary tests are often used to measure the potential or pre-illness level of intelligence. A somewhat related concern has led to some very interesting studies under such names as cognitive styles or cognitive control. See Gardner (1964) for an excellent introduction. Here the question is how a person uses his intelligence rather than how intelligent he is. Variables studied are, for instance, the degree of confidence a person has to have in his judgment before he will express it or act on it; whether more attention is given to rather global or more detailed aspects of a problem; how much judgment can be swayed by extraneous factors such as a distracting frame of reference (Witkin) or the opinions of others; and so on.

The second type of criticism really deals with the same facts but from a different point of view which can range from *statistical* concern with reliability to concern with the basic problems of measuring nonphysical variables. Such concern has led to the construction of scaling methods, which I believe have not been used much in the measurement of intelligence, and factor analysis, which is a method for the study of intercorrelations between a large number of variables.

Using factor analysis, items of the Binet were found quite early to be rather heterogeneous in nature (Jones, 1949). When the passes or failures of a number of subjects on Binet items were intercorrelated, the pattern of these correlations showed that a number of different types of tasks were present in the test and that these types were not represented equally at different age levels. When these groups of items are interpreted, it would be possible to select items on the basis of known functional unities, but until recently, these studies had little effect and did not affect the construction of new tests. For instance, while Wechsler's tests have a number of subtests, they were not selected on the basis of low intercorrelations and they do not represent independent abilities.

But things are changing; there are several attempts now to construct tests which will measure a number of separate abilities consistently throughout the age range for which they are designed. Meyers and Dingman (1960) constructed a preliminary set of tests for ages 2, 4, and 6, which give separate scores for motor speed, perceptual speed, language skills, reasoning, and memory, and they have separate materials for numerical ability. There is a Dutch battery, available in English translation, designed by Snijders and Snijders-Oomen (1958)—a husband-and-wife team—for children aged 5 to 16, which consists of eight parts: Mosaics, Picture Memory, Picture Arrangements, Picture Completion, Analogies, Knox Cubes, Copying Drawings, and Sorting. Factor analyses of baby and infant tests may lead to revisions of such well-known tests as the Bayley Scale and the Merrill Palmer Scale.

Besides the Binet and several other individually administered tests, there soon were what came to be called "paper-and-pencil" tests, which were administered to a group of persons at the same time. Almost as soon as several tests of intelligence existed, it was discovered that they did not correlate too well, either. Spearman suggested that whatever they had in common be called general intelligence or g. He also developed a statistical method to determine to what extent each test was measuring this g. What the test measures in addition to g, he called s for specific. The trouble with this proposal is, as Thurstone pointed out, that the addition or omission of each test changes the common nature of the set of tests and therefore of g.

Thurstone proposed instead a model in which specifics common to several tests were elevated to the position of independent co-equal factors. He thought that such factors would not be changed by the addition or omission of tests to the battery, as long as they did not concern this particular factor. Of course, the tests measuring a given factor could not be added to or subtracted from without changing that factor.

Most of the earlier studies of twins (and of foster children and their biological and foster parents) used single number measures of intelligence. Since these tests differ considerably in their composition, it is not too surprising to find different results when the influence of heredity was being studied.

In a study of twins at the University of Michigan and in further work at the University of Louisville Medical School, a variety of cognitive tests have been used in an attempt to see if some of these independent abilities are more under hereditary control than are others.

The first tests to be considered were the Chicago Primary Mental Abilities Tests constructed by L. L. Thurstone and Thelma G. Thurstone (1941). This battery had already been used by Blewett (1954) and by Thurstone, Thurstone, and Strandskov (1955). Because it was also used in the Michigan study (Vandenberg, 1962a) and in Louisville (Vandenberg, 1964), it is possible to present the results from four independent samples in Table 3.

TABLE 3

F-RATIOS OF DZ AND MZ WITHIN-PAIR VARIANCES FOR THE SIX SCORES
ON THE CHICAGO PRIMARY MENTAL ABILITIES TEST.

Name of Score	Blewett† (1954) $N_{DZ}26$ $N_{MZ}26$	Thurstone et al. (1955) $N_{DZ}53$ $N_{MZ}45$	Vandenberg (1962a) $N_{DZ}37$ $N_{MZ}45$	Vandenberg (1964) $N_{DZ}36$ $N_{MZ}76$
Verbal	3.13**	2.81**	2.65**	1.74**
Space	2.04*	4.19**	1.77*	3.51**
Number	1.07	1.52	2.58**	2.26**
Reasoning	2.78**	1.35	1.40	1.10
Word Fluency	2.78**	2.47**	2.57**	2.24**
Memory	not used	1.62*	1.26	not used

** indicates $p < .01$.
* indicates $p < .05$.

† F-ratios calculated from the h^2 values reported by Blewett by the formula $F = \dfrac{1}{1 - h^2}$.
This may result in minor inaccuracies that should not affect the general trend of the conclusions arrived at in the discussion.

The four studies agree on hereditary components in the verbal score and the word fluency score. These are independent factors in the general population, but is it possible that they are controlled by the same or an overlapping set of genes? There is also good agreement on the Spatial score. This is an ability to deal with two- and three-dimensional patterns in one's mind, which must be present in an unusual degree in Watson and Crick, the developers of a detailed three-dimensional model of the biochemical nature of the "genetic code."

Two studies also agree on the somewhat lower significance of hereditary factors in the memory score. The procedures employed to obtain this score require the subject to memorize a list of first and last names. The list of last names is then presented in a different order, and the subject has an answer sheet with a number of first names for each last name, from which he has to pick the correct one. The same procedure is followed for the second part in which the subject has to match two-digit numbers and objects as, for instance, 21 chair, 46 ball. It is clear that differences in the willingness to work hard at the memorizing greatly affect the results, so that these tests measure an unknown mixture of motivation and memory ability.

On the remaining two scores, the British study by Blewett disagrees with the American studies. Blewett found no evidence for an hereditary factor in the Number score, which is based on very simple arithmetic tests (adding four two-digit numbers, multiplying a two-digit number by a one-digit number, and underlining each number—on a page full of numbers—that is three higher than the preceding one). In Thurstone's study the results were not significant but in the direction of an hereditary component, while both groups of twins studied by Vandenberg gave significant evidence. With the Reasoning score the story is exactly reversed; Blewett's subjects gave evidence for an hereditary component, but none of the three American groups did.

One might speculate about differences between the groups in socioeconomic composition or in the educational practices to which they were exposed, but it would be good to keep in mind simpler explanations, such as sampling errors or differences in instructions.

In the edition used in the Michigan and Louisville studies, five of the six scores on the Primary Mental Abilities Test were obtained from three subtests each. It is thus possible to see whether the tests used to obtain each of these scores give different results as well as check on the consistency of such differences. These results are given in Table 4.

The results are rather consistent for the three verbal tests, quite consistent for the spatial tests and the reasoning tests, but jump around somewhat for the number tests and the word fluency test. In the spring of 1964 we concentrated on the spatial and number areas and administered a battery of thirteen spatial tests and seven short arithmetic tests, plus a few unrelated tests of vocabulary, perceptual speed, and reasoning. This will allow examination of the question whether all these spatial tests share a common hereditary component as well as give further information about the heritability of number ability.

Returning to the verbal tests for a moment, the finding that the multiple-choice vocabulary test shows no evidence of an hereditary component while the other two do is somewhat puzzling. In the Michigan study two more vocabulary tests were used: the orally administered vocabulary test of the Wechsler Test and an abbreviated form of the Ammons Picture Vocabulary Test, in which the subject is shown a series of cards with four pictures and is asked to indicate the correct picture for a number of words. The F-ratio for the Wechsler Vocabulary was 2.19* and for the Ammons Picture Vocabulary 1.42. Considering the results on all verbal tests, it appears as if mere recognition of words is not enough to bring out hereditary differences, since the tests that have higher F-ratios require either independent production (Wechsler) or some searching (PMA completion and sentences).

A number of other tests have been screened, both in Louisville and earlier in the Michigan study, for their promise in twin studies. Table 5 summarizes those

TABLE 4

F-RATIOS OF DZ AND MZ WITHIN-PAIR VARIANCES FOR EACH OF
THE SEPARATE SUBTESTS OF THE PRIMARY MENTAL ABILITIES TEST BATTERY.

Name of Test	Michigan $N_{DZ}37 \ N_{MZ}45$	Louisville $N_{DZ}36 \ N_{MZ}76$
Number		
Addition	2.40**	1.34
Multiplication	1.75*	3.49**
Three Higher	2.05*	2.38**
Verbal		
Sentences	1.80*	2.09**
Vocabulary	.99	.97
Completion	3.01**	1.83*
Spatial		
Flags	1.80*	1.94**
Figures	1.79*	1.94**
Cards	1.77*	1.84*
Word Fluency		
First letters	1.90*	1.44
Four-letter words	1.59	1.86*
Suffixes	2.31**	1.54
Reasoning		
Letter series	1.08	1.07
Letter grouping	1.78*	1.56*
Pedigrees	1.28	.93
Memory		
First names	1.57	not used
Word-number	1.05	not used

** indicates $p < .01$.
* indicates $p < .05$.

findings, although I will not attempt to discuss all of them in detail. Discrepancies between tests that appear, at a first glance, to be measuring the same thing may be of heuristic value to help us to understand better just which aspect of a test brings out hereditary differences.

Of the tests on this list I will call your attention only to the Gray Oral Reading Paragraphs, which examines such a highly practiced skill as reading and appears to have an hereditary component, similarly, at the bottom of the list, for clerical checking, spelling, and recognition of incorrect grammar.

TABLE 5

F-RATIOS OF DZ AND MZ WITHIN-PAIR VARIANCES FOR VARIOUS COGNITIVE TESTS.

	Vandenberg (1962a)		
	F	N_{DZ}	N_{MZ}
Gray Oral Reading Paragraphs	1.97*	36	45
DAT Spatial Reasoning	1.02	34	44
DuBois Object Aperture, Form A	1.21	34	45
Miller Mechanical Insight	1.44	34	45
Raven Progressive Matrices	1.80*	33	43
Arthur Stencils, Form 1	1.34	34	42
Alexander Passalong Test	1.16	32	38
Coxe Cube Construction	1.50	35	41
Coxe Cube Construction	1.27	35	41
Heim, AH4, Part 1	2.20**	34	45
Heim, AH4, Part 2	1.56	34	45
Hunt Memory for Faces	.98	34	45
WISC Memory for Digits	1.85*	31	42
Figure Classification (Thurstone)	1.58	34	43
DAT Verbal Reasoning	2.29**	25	47
DAT Numerical Ability	1.39	25	47
DAT Abstract Reasoning	1.47	25	47
DAT Spatial Reasoning	1.67	25	47
DAT Mechanical Reasoning	1.36	25	47
DAT Clerical Checking	2.54**	25	47
DAT Spelling	3.64**	25	47
DAT Grammar	3.06**	25	47

** indicates $p < .01$.
 * indicates $p < .05$.

Two tests that were constructed as measures of general intelligence—that is, the Raven Progressive Matrices and the AH4 test of Alice Heim—do show some hereditary variance but not as high as did tests of more specific abilities.

In a series of studies since the last war Guilford has administered batteries of new tests to rather select groups of adult subjects, often military personnel with fairly high rank. A good deal of thought and effort went into the construction of these new tests in which an attempt was made to get at problem solving and reasoning at a high level. In some of the tests the subject is not given a choice of answers but must write his own. Guilford calls such tests measures of "divergent thinking" as opposed to convergent thinking. They have frequently been used to measure creativity. Table 6 gives some results with twins for nine such tests. While the sample size of the fraternals is disappointingly low, the results show rather consistent lack of significance for a genetic factor. This matter deserves further study however, since the word-fluency score in the PMA battery, which also calls for divergent thinking, did reach significance as earlier reported. Unfortunately, the evaluation of the written answers is quite time consuming, especially when more than one judge is used.

TABLE 6

F-RATIOS OF 24 DZ AND 67 MZ WITHIN-PAIR VARIANCES ON
NINE DIVERGENT THINKING TESTS.

Guilford Tests of Divergent Thinking	
1. Pertinent Questions	1.85*
2. Different Uses	1.53
3. Social Institutions	1.39
4. Seeing Deficiencies	1.35
5. Making-a-Plan	1.11
6. Similar Words	1.10
7. Associations	1.08
8. Figure Production	1.03
9. Picture Arrangement	.94

* indicates $p < .05$.

I have information about still other tests, but by now the thought must have occurred to you that it might be possible that most of these tests are measuring the same thing genetically, even if the correlations between them are low when ordinary subjects are studied. I use the word "ordinary" here in contrast with twins. That thought also occurred to me and I looked for a method that would allow us to test whether two variables are controlled by the same or partially overlapping genes.

My first thought was that the correlations between the twin differences would provide the answer. Looking first only at identical twins, differences on test one and on test two are due only to environment. If there is a correlation between these differences, it shows that the environment affects the two variables in a similar way. One could ask, for instance, whether the taller members of a number of identical twin pairs also are more dominant compared to their partners. One could also factor analyze a whole matrix of such correlations to see how many dimensions exist in the environmental influences.

When we turn to fraternal twins, differences between partners are due to heredity as well as environment. Similar action on several variables of heredity as well as of environment would enhance the correlations, so that the difference correlations for fraternal twins will generally be different from those found for identical

TABLE 7

Correlations of within-pair differences for

PMA Score	45 identical pairs						37 fraternal pairs					
	N	V	S	W	R	M	N	V	S	W	R	M
Number	*	−124	−187	−183	179	052	*	272[1]	454[2]	393[2]	461[2]	189
Verbal	−124	*	137	394[2]	289[1]	065	272[1]	*	215	521[2]	515[2]	239
Space	−187	137	*	281[1]	−102	−353[2]	454[2]	215	*	420[2]	472[2]	359[1]
Word Fluency	−183	394[2]	281[1]	*	438[2]	123	393[2]	521[2]	420[2]	*	429[2]	345[1]
Reasoning	179	289[1]	−102	438[2]	*	317[1]	461[2]	515[2]	472[2]	429[2]	*	308[1]
Memory	052	065	−353[2]	123	317[1]	*	189	239	359[1]	345[1]	308[1]	*
Number of Significant Correlations	0	2	2	3	3	2	4	3	4	5	5	3

[1] correlation significantly different from zero at 5% probability.
[2] correlation significantly different from zero at 1% probability.

twins. An example is given in Table 7, where difference correlations are given for 45 identical pairs and 37 fraternal pairs on the six scores of the Primary Mental Abilities Tests.

In these correlations the effect of age is largely eliminated when twin differences are studied, unless older twins generally differ more from one another than do young ones, in which case higher correlations between twin differences on two tests might be expected for older twins than for younger pairs, if two tests measure related phenomena. In the age range of 12 to 18 years, this effect of age was probably minimal.

Looking first at the results for the identical twins in which differences are only due to the environment, whether prenatal or postnatal, we see that of the fifteen different correlations, only three are significantly different from zero at the 5 per cent level of significance and three additional ones at the 1 per cent level. Differences in Number ability are not correlated with differences in the other five scores. Verbal ability—that is, vocabulary knowledge as demonstrated in multiple-choice tests—is correlated with word fluency scores. (These are based on the number of words with a given first letter or ending which the subject could write down in a few minutes.) The latter score is also correlated with the Reasoning score. The spatial visualization or Space score correlates negatively with the Memory score. The three 5 per cent level correlations will not be considered. It appears then that within these identical twin pairs and for these six variables, the environmental influences have largely independent effects. It may be worthwhile to employ this method with more variables to study the relations between influences thought to produce differences between siblings within a family, such as birth order, size of family, and so forth.

When we now turn to the results for the fraternal twins, the story is quite different. Twelve of the fifteen correlations are significant, eight at the 1 per cent and four at the 5 per cent level of significance. The only correlations that are not significant are between Number ability and Verbal abilities, between Memory and Space, and between Memory and Word Fluency. The correlations still are not high, accounting at most for one fourth of the variance, but with two exceptions they are higher than the correlations for the identical twin differences. If we simplify matters grossly and assume that the environmental contribution to the within-pair variance is the same for the fraternals as for the identicals and subtract the identical difference correlations from the fraternal ones (after Fisher's z-transformation) we obtain the values shown in Table 8 for the hypothetical correlations for fraternal within-pair differences due to heredity only.

While many objections can be raised to this procedure it makes possible the simultaneous grasp of the two sets of correlations and thus facilitates some interpretation. Actually only four of the fifteen correlations differ at a statistically significant level from one matrix to the other. These four are starred in Table 8.

Let us look now at the obtained values and interpret them as estimates of correlations such as would be obtained between that part of fraternal twin differences that is due to hereditary factors on one test and the "hereditary" difference scores

TABLE 8

HYPOTHETICAL CORRELATIONS OF TWIN DIFFERENCES DUE TO HEREDITY ONLY.

	N	V	S	W	R	M
Number	—	38	59*	54*	31	19
Verbal	38	—	08	16	27	18
Space	59*	08	—	16	55*	63*
Word fluency	54*	16	16	—	—01	23
Reasoning	31	27	55*	—01	—	—01
Memory	19	18	63*	23	—01	—

* indicates that the original two correlations differed significantly, that is, $r_{DZ} \neq r_{MZ}$.

on another test, if these could be measured directly. There appears to be some relation between Number ability on the one hand and Space and Word fluency on the other (but not between these two) and also between Space on the one hand and Reasoning and Memory on the other hand (but not between these two either). The hypothetical correlations are low and would indicate near independence of all six abilities. But enough has been made of this crude analysis.

The following method suggested itself while searching for a more acceptable method to answer the question whether these six abilities were genetically related. Consider the original F-tests for the statistical significance of the increase in within-pair variance of fraternal twins as compared with identical twins. For each separate variable we calculate the ratio:

$$F = S^2_{wDZ} / S^2_{wMZ} \qquad (1)$$

and evaluate this for N_{DZ} and N_{MZ} degrees of freedom, where S^2_w is the within-pair variance of N_{DZ} and N_{MZ} pairs of twins A and B on test x.

$$S^2_w = \frac{1}{2N} \sum (x_A - x_B)^2 \qquad (2)$$

Now we wish to generalize the F-test to a multivariate comparison. Let the within-pair covariances for scores i and j

$$Cov_{wij} = \frac{1}{2N} \sum (x_{iA} - x_{iB})(x_{jA} - x_{jB}) \qquad (3)$$

form the elements of the covariance matrices C_{DZ} and C_{MZ}. Then the generalization of formula (1) would be: how many significant roots has the characteristic equation

$$| C_{DZ} - \Lambda C_{MZ} | = 0 \qquad (4)$$

A possible interpretation of this equation might be: given C_{MZ}, the within-pair covariance matrix of scores for the monozygous twins, find the (asymmetric) matrix of eigenvectors which will transform C_{MZ} into C_{DZ} (the dizygous within-pair covariance matrix), where this asymmetric transformation matrix is thought of as representing the addition of hereditary differences.

Equation (4) was solved for the six PMA scores, and the six roots shown in

Table 9 were obtained. To test the significance of these roots Bartlett's (1950) chi-squared test of the homogeneity of the remaining roots after extraction of 1, 2, . . . , k roots was applied. The results are shown in Table 10 and indicate that the probability is less than .05 that fewer than four roots are significant, or in other words, that in this sample there are at least four independently significant hereditary components.

TABLE 9

Roots of $\mid C_{DZ} - \Lambda\, C_{MZ} \mid = 0$	
1	3.99972
2	2.23172
3	1.58655
4	1.00141
5	.64598
6	.38206

TABLE 10
CHI-SQUARED VALUES FOR THE HOMOGENEITY OF
THE REMAINING ROOTS AFTER EXTRACTION OF K ROOTS.

k	chi square	d.f.	p
1	30.698	20	between .10 and .05
2	17.483	6	less than .01
3	7.250	2	less than .05
4	2.147	5	n.s.
5	0	0	

It is of interest to compare these roots with the F-ratios for the DZ over the MZ within-pair variances for the six original scores. Table 11 shows these F-ratios; four of them are significant beyond the 1 per cent level.

TABLE 11
F-RATIOS OF FRATERNAL OVER IDENTICAL WITHIN-PAIR VARIANCES
OF SIX PMA SCORES FOR 37 AND 45 DEGREES OF FREEDOM.

	F	p
Number	2.583	.01
Verbal	2.651	.01
Space	2.419	.01
Word Fluency	2.479	.01
Reasoning	1.401	n.s.
Memory	1.213	n.s.

We concluded that there are at least four independent hereditary components in the six PMA scores of the subjects in this sample. Since the analysis of the single scores led to the conclusion that Number, Verbal, Space, and Word

Fluency were under a statistically significant degree of genetic control it may be warranted to conclude that the four independent components represented by the significant roots of $| C_{DZ} - \Lambda C_{MZ} | = 0$ are rather similar to the Number, Verbal, Space, and Word Fluency abilities.

For this symposium I wanted to have another example of the use of this multivariate method, this time of a situation in which I expected that the variables would share more common hereditary determination. I believed that this might be the case with the subtests of the WISC. We had collected results on the Wechsler Intelligence Scale for Children on 64 pairs of twins, 5 to 14 years of age. When the results of the bloodtyping were received it turned out that there were 48 pairs of identical, but only 16 pairs of like-sexed fraternal twins. Although this is not really adequate for a definitive analysis, I was, at the time when this paper was being prepared, curious enough to proceed with a preliminary analysis. Table 12 shows that only four out of the eleven subtests gave evidence of an hereditary factor at the 5 per cent level. The analysis will have to be repeated on a larger group before these results are firmly accepted.

TABLE 12

F-RATIOS OF DZ AND MZ WITHIN-PAIR VARIABLES FOR THE WISC.

Name of Subtest	F	N_{DZ}	N_{MZ}
Information	2.00*	16	48
Comprehension	1.46	16	48
Arithmetic	1.63	16	48
Similarities	.74	16	47
Vocabulary	1.99*	16	48
Digit Span	1.37	10[a]	32[a]
Picture Completion	1.38	16	48
Picture Arrangement	2.30*	16	48
Kohs Block Designs	.75	16	48
Object Assembly	.87	16	47
Digit Symbol Substitution	1.86*	15[a]	44[a]
Verbal IQ	2.34*	16	48
Performance IQ	1.91*	16	48
Total IQ	2.43**	16	48

** $p < .01$.
* $p < .05$.
[a] Time limitations necessitated occasional elimination of these subtests.

The correlations between the twin differences on the eleven WISC subtests are shown on the left in Table 13, separately for the dizygous and the monozygous twins. On the right are shown those principal factors in these correlation matrices which had significant roots, after they had been rotated according to the Varimax criterion. Also shown are communalities, that is, the percentages of the variance for each test which is accounted for by these factors. It is clear that

TABLE 13

CORRELATIONS BETWEEN DIFFERENCES ON ELEVEN WISC SUBTESTS AND PRINCIPLE AXIS FACTORS IN THESE CORRELATIONS, AFTER VARIMAX ROTATION, FOR DZ AND MZ TWIN PAIRS.

Difference correlations for 16 DZ twin pairs

WISC Subtest	Info	Comp	Arith	Simil	Vocab	D.Span	P.Comp	P.Arr	Kohs	Obj.A	Sym.C	Varimax Rotation							h^2
Information												93	00	18	03	-08	16	07	93
Comprehension	74											72	-25	30	16	-12	12	-34	83
Arithmetic	49	45										34	00	80	37	06	-01	16	93
Similarities	-22	-22	-34									-19	-66	-35	-04	20	-07	29	73
Vocabulary	51	41	25	-06								39	-16	13	04	20	74	-03	78
Digit Span	24	34	68	-32	30							11	05	85	31	14	-17		89
Picture Completion	-12	-07	25	04	26	51						-11	05	26	11	86	11	-05	85
Picture Arrangement	07	07	06	-48	-18	-20	-53					05	45	-10	26	-62	-02	-26	74
Kohs Blocks	-22	-45	-12	-46	-36	-06	11	33				-23	87	-06	00	05	-13	12	85
Object Assembly	-31	-36	-24	01	02	05	-15	-08	21			-40	12	08	-49	-18	33	25	63
Symbol Coding	09	25	46	-10	10	-01	14	26	04	-44		05	05	11	92	00	04	-01	87

Difference correlations for 48 MZ twin pairs

WISC Subtest	Info	Comp	Arith	Simil	Vocab	D.Span	P.Comp	P.Arr	Kohs	Obj.A	Sym.C	Varimax Rotation			h^2
Information												03	12	-04	02
Comprehension	11											08	71	-01	51
Arithmetic	-03	26										64	16	-02	44
Similarities	01	19	32									39	21	-06	20
Vocabulary	04	34	34	20								54	-15	-09	32
Digit Span	06	20	34	20	06							36	-19	25	23
Picture Completion	23	17	-14	-11	15	-01						-19	18	30	16
Picture Arrangement	-03	15	03	-01	-17	15	-08					09	-42	-12	20
Kohs Blocks	-16	03	-05	04	-17	15	-09	-08				-05	04	52	27
Object Assembly	-14	-05	-17	-02	-16	04	20	15	20			-19	00	34	15
Symbol Coding	05	10	-15	12	02	09	08	-01	-09	10		08	-28	33	19

the correlations between the MZ twin differences on the eleven WISC subtests are quite small and that the common within-pair environmental factors account for only a small proportion of the variance. The story is different for the DZ difference correlations. Their values are frequently much higher and the common factors account for a major proportion of the variance. There are, however, a rather large number of factors, which might indicate that these tests measure a number of different hereditary components if we assume that the within-pair environmental effects would be the same. The sample size is too low to have any confidence in these results.

It is of interest to see whether the differences between the members of MZ twin pairs are related to some family attributes. Correlations for the differences on the 11 WISC subtests and the total WISC score (the "IQ") and the following ten variables were obtained:

1. The age of the twins when tested. One would expect a larger twin difference, the older the pair is.

2. A numerical rating of the father's occupation based on tables in a book entitled *Occupations and Social Status* by Reiss (1961). A higher socioeconomic status might provide more opportunity for individual differences, both because of material conditions and family attitudes.

3. and 4. The number of years of education of the father and the number of years of education of the mother. Same hypothesis.

5. The total number of children in the family. Larger families might provide more opportunities for social contacts and learning but less time for individual treatment of twins.

6. and 7. The age of the father when the twins were born and the age of the mother when the twins were born. Younger parents might devote more time to their children and might foster individual difference between twins more.

8. Present income of the family. This was estimated from the family's home address by the use of a study of the 132 census tracts of Louisville and Jefferson County, Kentucky, and of 22 census tracts in Clark and Floyd counties in Indiana in which median incomes for each tract are reported (Louisville Chamber of Commerce, 1962).

9. The sex of the twins. Girls were coded 1 and boys 2, so that a negative correlation indicates that girls had larger twin differences.

10. The number of children preceding the birth of the twins. Perhaps later-born children receive more stimulation from their brothers and/or sisters, but less from their parents.

The 120 correlations are shown in Table 14. Only six of them are significant at the .05 level of probability, which is not much more than chance. For that reason I will abstain from interpretation. I hope to repeat this with a larger sample and perhaps with more sensitive measures.

TABLE 14

Correlations between family variables and WISC subtest differences for 48 MZ twin pairs.

Family Variables	Information	Comprehension	Arithmetic	Similarities	Vocabulary	Digit-Span	Picture Completion	Picture Arrangement	Kohs Blocks	Object Assembly	Symbol Coding	WISC IQ
1. Age of twins	12	34*	24	25	17	18	07	16	20	18	04	24
2. Father's occupation	09	10	-03	13	-06	05	-06	02	12	-06	05	10
3. Education of father	19	-02	-01	10	-13	-07	13	06	-03	03	04	13
4. Education of mother	-03	-14	-08	19	-16	-02	-11	04	-16	-01	-04	-12
5. Size of family	19	27	10	07	13	-12	01	-06	-19	06	-33*	11
6. Father's age } at birth	20	15	-11	01	12	12	16	-19	11	13	04	08
7. Mother's age } of twins	14	30*	02	15	16	07	12	-29*	15	13	-08	16
8. Present income	-04	08	02	34*	-05	-27	28*	-03	09	-06	10	21
9. Sex of twins	-20	-03	07	09	07	05	-01	-16	03	-08	09	-06
10. Parity of twin births	11	20	-06	-08	01	-06	25	-19	09	16	-17	19

WISC subtests

$* \ p < .05.$

TEST-TAKING ACCURACY

Before leaving the topic of intelligence to turn to my final section, which will deal with personality, I should like to mention briefly an attempt to study accuracy or rather the tendency to make mistakes.

In a number of the tests we used, only a few alternative answers are provided, so that the subject could get some answers right by mere guessing. In such tests there is a "correction for guessing"; in other words, to obtain the test score a certain percentage of his wrong answers is subtracted from his correct answers. The time limits for these tests are short so that very few people are able to attempt all items. The number of wrong answers is therefore not simply the complement of the number of right answers. Table 15 shows that there were instances in which the heritability of the number of wrong answers was greater than the heritability of the number right.

To see if perhaps we were measuring a general test-taking accuracy, intercorrelations were calculated between the number of mistakes on thirteen paper-and-pencil tests and the variability of the performance on the Rod and Frame Test which will be described below in the section on personality. Many of the correlations between the number of mistakes on these thirteen paper-and-pencil tests for all the high school students were significant, although not very high. There were no correlations with the Rod and Frame Test. The correlations are shown in Appendix 1. The same computer program that calculated these correlations also computed correlations between the twin differences on these fourteen variables, separately for the DZ and for the MZ twins. These are shown in Appendix 2 and Appendix 3. Table 16 shows the distributions of these three sets of correlations. The MZ twin difference correlations center around the .00 to .14 interval; it appears that the within-pair environmental influences that lead to twin differences in the tendency to make mistakes do not have very general effects. The DZ twin difference correlations are considerably higher and their distribution looks more like the distribution of the values of the ordinary correlations. It is tempting to conclude that these tests brought out one or several general hereditary components related to a tendency to make mistakes, or perhaps the other way around, to a tendency to be accurate.

The possibility of obtaining a general measure of accuracy or inaccuracy in taking tests was discussed earlier by Fruchter (1950), who concluded that this is quite feasible and that a measure of such an attribute is independent of the abilities measured by these tests.

In future studies, care should be taken that the tests contain enough questions and the time limits are short enough, so that no one finishes, so that the number right and the number wrong are experimentally independent, even though, of course, a tendency to be hasty and inaccurate would lower the ability score, and ability would lower the inaccuracy score. In later studies it may be worthwhile to estimate each of these scores by regression techniques in such a way that the correlation due to this logical relationship is removed. The use of separate sheets for recording the answers may well raise general inaccuracy and is not recommended with high school students for studies of this kind. In the tests we used, the test was generally its own answer sheet.

TABLE 15

COMPARISON OF HERITABILITIES FOR THE CORRECT SCORE AND
FOR THE NUMBER OF WRONG ANSWERS ON A VARIETY OF TESTS.

Name of Test	F-ratio for	
	correct score	wrong answers
PMA Number Test		
Addition	2.40**	.76
Multiplication	1.75*	.90
Three higher	2.05*	5.04**
PMA Verbal Test		
Sentences	1.80*	1.66
Vocabulary	1.00	3.04**
Completion	3.01**	1.81*
PMA Spatial Test		
Flags	1.80*	2.93**
Figures	1.79*	2.39**
Cards	1.77*	1.45
PMA Reasoning Test		
Letter series	1.08	.65
Letter grouping	1.78*	1.71*
Pedigrees	1.29	.83
PMA Memory Test		
First names	1.57	1.14
Word-number	1.05	1.03
Concealed Figures	3.05**	1.88*
Reading Mirror Typing	1.84*	1.80*
McGill Closure Test	2.31**	1.19
Hidden 4-letter words	2.10**	.46
Copying	1.48	2.03*
Gestalt Completion, form A	1.63	1.24
form B	2.33**	.98
Identical Forms	1.60	1.64
Figure Classification	1.55	1.35
Object Aperture	1.21	1.02
DAT Space Relations	1.02	1.19
Ammons Picture Vocabulary Test	1.14	.99
Memory for Faces	.98	1.10

** indicates $p < .01$.
* indicates $p < .05$.

TABLE 16

CORRELATIONS FOR 164 HIGH SCHOOL STUDENTS OF THE NUMBER OF
WRONG ANSWERS ON THIRTEEN PAPER-AND-PENCIL TESTS AND THE VARIABILITY ON
THE ROD AND FRAME TEST (ON THE RIGHT) AND TWIN DIFFERENCE CORRELATIONS
ON THESE VARIABLES FOR 37 DZ AND 45 MZ PAIRS.

Range of Values of the correlations	Distribution of correlations* between		
	37 DZ twin differences	45 MZ twin differences	values for all 164 students
.60 to .74	7	0	3
.45 to .59	24	2	11
.30 to .44	18	10	35
.15 to .29	24	19	22
.00 to .14	16	28	18
−.15 to −.01	1	20	2
−.30 to −.14	1	10	0
−.45 to −.29	0	2	0

* Because only the correlations below the leading diagonal are counted, the actual frequencies are double the size shown here.

PERSONALITY

This discussion of findings on genetic factors in personality variables is divided into temperament, ANS reactions, attitudes and interests, and values.

HEREDITARY DIFFERENCES IN TEMPERAMENT

Most personality questionnaires contain items concerned with all of these categories, even the second one, for instance when the question is asked whether or how often one blushes, how one feels after a shocking occurrence, and so forth. Here the term will be used in connection with personality questionnaires, because most of these aim at measuring the more enduring underlying personality rather than attitudes, interest, and values, which may perhaps be more changeable.

The only personality questionnaire on which there are data from more than one twin study is Cattell's High School Personality Questionnaire (Cattell *et al.*, 1958). There are in fact three studies, but unfortunately a different edition of this questionnaire was used in each study (Cattell *et al.*, 1955; Vandenberg, 1962; Gottesman, 1963) so that lack of replication may be due to changes in the questionnaire or to differences between the groups or to unstable estimates of low heritability. Since in all three studies the statistical significance was low or only at the .05 level for most parts of this questionnaire, the latter may be the most serious source of the lack of agreement.

In the same study Gottesman (1963) also reported that three of the ten scales of the Minnesota Multiphasic Personality Inventory (MMPI) gave evidence at the .05 level of an hereditary component. These scales are D (depression), Pd (psychopathic personality), and Si (social introversion). Two more scales approached the .05 level: Pt (psychasthenia) and Sc (schizophrenia).

The MMPI was developed by contrasting the responses of various groups of neuropsychiatric patients with the responses of normals and retaining questions with widely differing frequencies of yes and no responses. While the MMPI has not been as successful in differential diagnosis as was hoped, it has been found to correlate with various criteria.

Recently we administered the Myers-Briggs Type Indicator to 27 pairs of fraternal and 40 pairs of identical twins. This questionnaire was designed to determine the subject's position with respect to the typology of Jung, who described people as being extrovert (E) or introvert (I); more sensation oriented (S) or more intuitive (N); more thinking (T) or more feeling (F); and more judging (J) or more perceiving (P). The F-ratios for these four bipolar scales were E-I, 1.84; S-N, .70; T-F, .80; and J-P, .76. Here is further evidence that introversion may be in part genetically determined, even though Jung's concept of introversion did not have the connotation of neuroticism which it generally has acquired in psychology (Myers, 1962).

The state of the art of personality questionnaires will undoubtedly improve now that test scoring by optical scanning and item analysis and correlational studies by computers can be used to cut down on the tedious part of test construction and validation. Correlations between some of the scales of different questionnaires are fairly high, even if the names are occasionally different, so that it appears that these questionnaires are on the right track.

For twin studies, and later for family studies, special tests may have to be constructed, aimed especially at hereditary variation in personality. Loehlin (1965) has proposed a comparison of the DZ and MZ concordance separately on each question of personality tests followed by correlational analysis. This method would allow us to construct tests that would be more saturated with items tapping hereditary differences.

There have been many attempts to measure differences in personality by more objective methods, in which the subject is asked to perform some task. In fact, there have been too many to attempt even a very condensed review here. I will only mention the Rod and Frame Test, which is concerned with the accuracy and variability of the perception of the upright in a conflictful situation. The subject sees, in a darkened room, a luminescent rod inside a square frame. With the frame in a tilted position, the subject is asked to put the rod in the true vertical position. The tilted "frame of reference" has been found to affect the performance of different individuals to a varying degree. The subject may, in addition, be seated in a tilted chair.

Witkin and his colleagues (1954) have proposed an interesting personality dimension that they call field dependence, which they describe as the degree to which a person is autonomous or more dependent on continual reinforcement from the outside. This is somewhat reminiscent of the distinction between inner-directed and other-directed individuals made by Riesman (1950).

In the Michigan study I (1962a) obtained an F-ratio of 1.69 for the comparison of the DZ and MZ within-pair variances, indicating a considerable hereditary component in this variable.

ANS Functions

There have been few studies of autonomic nervous system functions in twins. Vandenberg, Clark, and Samuels (1965) found evidence for an hereditary component in the changes in heart rate and breathing rate after unexpected light flashes or noise, but not in changes in the galvanic skin response. Kryshova and her colleagues (1963) in Leningrad reported higher concordance for MZ than for DZ twins in changes in blood pressure or skin temperature after touching the skin with hot or cold objects as well as for changes in blood pressure after (unexpected?) stimulation with bell or light.

This area seems especially promising now that newer equipment can provide "on line" data analysis, and eliminate the tedious work of counting or measuring marks on rolls of paper produced by a polygraph.

Attitudes and Interests

Contrary to what one might expect, attitudes and interests may have a rather stable anchoring in hereditary factors. At least this is what one would conclude from Carter's (1932) finding of greatly increased concordance of identical twins (compared to fraternal twins) on the Strong Vocational Interest Blank. This test, constructed with great care and found to have unusually high reliability and validity (Strong, 1955), was used for a replication of Carter's study by Vandenberg and Kelly (1964). In both studies the high heritability estimates tended to cluster in the "Science" group of occupational interests. It is quite likely that ability helps to determine the responses to the questions in this vocational interest questionnaire, even though one customarily warns subjects that this "test" does not measure whether one is qualified for a given occupation, but only helps to find out the kind of occupation that would suit one's interests and likes and dislikes. It is, in fact, extremely likely that we will find that current distinctions between personality and cognition, motor skills and perception will be meaningless at the level of the genes. The groupings will probably be quite different.

Values

Only one twin study (Nolan, 1958) used this most culturally determined attribute of personality. Nolan administered the Allport-Vernon-Lindzey Study of Values Scale to twelve pairs of MZ and ten pairs of DZ twins and obtained data from which I calculated the F-ratios shown in Table 17. While the sample size is too small for significance, the results are suggestive.

It is at least interesting to speculate whether the rather different orientations of the major human cultures were in part determined by differences in gene frequencies rather than being entirely due to environmental (historical, geographical, and so forth) factors.

TABLE 17
F-RATIOS OF DZ AND MZ WITHIN-PAIR VARIANCES
FOR SIX VALUE ORIENTATIONS.

Predominant Interest	F
Theoretical	1.34
Economic	2.02
Esthetic	2.58
Social	1.94
Political	3.09
Religious	2.39

CONCLUDING REMARKS

There is a growing interest in behavior genetics, after a period in which social scientists showed very little interest in hereditary differences, or even denied their importance.

New statistical techniques have been proposed by Cattell (1953), Guttman and Guttman (1963), Loehlin (1965), Nichols (1965), and Vandenberg (1965). We may hope for larger samples and frequent replication of studies as more adequate statistical methods are used. Before long, enough may be known about some psychological variables to warrant the expense of a family study.

As more and more twins are being bloodtyped, it becomes possible to start looking for linkage of the blood groups with quantitative variables. It would be highly desirable if all twin studies obtained a complete set of blood tests on discordant twins for that purpose. They should also include a measure of height, since this might be easiest to obtain and since it would provide a good model to work on if sufficient data can be pooled. As a matter of fact, anthropometric twin studies may very well provide good material on which to try out some of the newer statistical methods, because such variables are less subject to error or culturally determined differences between samples and easier to visualize concretely, especially in multivariate analyses.

There appears to be a good deal of promise for future studies of human behavior genetics, although until now no fundamentally new findings have been developed; however, older findings have been verified with somewhat improved techniques.

APPENDIX 1

Correlations between the number of mistakes on thirteen paper-and-pencil tests and the Rod and Frame Test for 164 high school students.

	1	2	3	4	5	6	7	8	9	10	11	12	13	14	
	*	63	54	29	13	23	22	34	13	35	48	31	26	00	(1) Mutilated words, form A
	63	*	33	39	18	30	24	33	13	33	42	41	34	−04	(2) Mutilated words, form B
	54	33	*	17	17	25	30	25	10	36	39	29	22	00	(3) Incomplete words
	29	39	17	*	43	42	41	35	06	56	44	38	31	03	(4) Letter series
	13	18	17	43	*	36	49	14	07	45	27	34	36	06	(5) Mechanical insight
	23	30	25	42	36	*	54	28	36	47	40	32	36	02	(6) AH4 test, Part 1
	22	24	30	41	49	54	*	18	22	48	41	44	53	04	(7) AH4 test, Part 2
	34	33	25	35	14	28	18	*	16	35	37	16	16	−01	(8) DAT Spelling Test
	13	13	10	06	07	36	22	16	*	15	19	10	33	11	(9) PMA N_3
	35	33	36	56	45	47	48	35	15	*	61	44	52	04	(10) PMA V_2
	48	42	39	44	27	40	41	37	19	61	*	46	44	08	(11) PMA V_3
	31	41	29	38	34	32	44	16	10	44	46	*	62	15	(12) PMA S_1
	26	34	22	31	36	36	53	16	33	52	44	62	*	03	(13) PMA S_2
	00	−04	00	03	06	02	04	−01	11	04	18	15	03	*	(14) Rod and Frame Test

Correlations of .20 are signicant at $p < .01$.

126

TWIN DIFFERENCE CORRELATIONS BETWEEN THE NUMBER OF MISTAKES
ON THIRTEEN PAPER-AND-PENCIL TESTS AND THE VARIABILITY ON
THE ROD AND FRAME TEST FOR 37 DZ TWIN PAIRS.

	1	2	3	4	5	6	7	8	9	10	11	12	13	14
(1) Mutilated words, form A	*	63	55	35	11	27	43	36	20	50	47	47	28	06
(2) Mutilated words, form B	63	*	21	31	12	28	37	20	29	20	29	49	23	01
(3) Incomplete words	55	21	*	16	17	27	50	30	12	40	50	48	38	00
(4) Letter series	35	31	16	*	35	30	48	33	03	70	48	43	33	04
(5) Mechanical insight	11	12	17	35	*	27	14	29	−16	16	26	22	10	20
(6) AH4 test, Part 1	27	28	27	30	27	*	61	55	39	43	48	45	38	11
(7) AH4 test, Part 2	43	37	50	48	14	61	*	62	37	47	57	62	55	−03
(8) DAT Spelling Test	36	20	30	33	29	55	62	*	21	46	68	49	54	03
(9) PMA N_3	20	29	12	03	−16	39	37	21	*	14	27	20	37	22
(10) PMA V_2	50	20	40	70	16	43	47	46	14	*	56	50	57	02
(11) PMA V_3	47	29	50	48	26	48	57	68	27	56	*	50	52	15
(12) PMA S_1	47	49	48	43	22	45	62	49	20	50	50	*	71	03
(13) PMA S_2	28	23	38	33	10	38	55	54	37	57	52	71	*	03
(14) Rod and Frame Test	06	01	00	04	20	11	−03	03	22	02	15	03	03	*

Correlations of .41 are significantly different from zero
at $p < .01$; correlations of .32 at $p < .05$.

APPENDIX 3

TWIN DIFFERENCE CORRELATIONS BETWEEN THE NUMBER OF MISTAKES ON THIRTEEN PAPER-AND-PENCIL TESTS AND THE VARIABILITY ON THE ROD AND FRAME TEST FOR 45 MZ TWIN PAIRS.

	1	2	3	4	5	6	7	8	9	10	11	12	13	14
(1) Mutilated words, form A	*	34	25	24	-07	04	-07	07	-07	02	29	08	-07	00
(2) Mutilated words, form B	34	*	47	25	10	06	23	12	-21	18	33	-08	12	-14
(3) Incomplete worsd	25	47	*	29	-06	03	-13	31	-08	11	26	04	-29	-19
(4) Letter series	24	25	29	*	15	14	04	27	10	-02	39	11	-23	-09
(5) Mechanical insight	-07	10	-06	15	*	31	30	10	23	28	-06	-39	21	-22
(6) AH4 test, Part 1	04	06	03	14	31	*	24	14	29	42	-12	-42	-05	-07
(7) AH4 test, Part 2	-07	23	-13	04	30	24	*	-03	11	11	06	-20	25	03
(8) DAT spelling Test	07	12	31	27	10	14	-03	*	17	31	39	02	00	-01
(9) PMA N_3	-07	-21	-08	10	23	29	11	17	*	-06	-07	-27	-08	-20
(10) PMA V_2	02	18	11	-02	28	42	11	31	-06	*	30	-19	14	18
(11) PMA V_3	29	33	26	39	-06	-12	06	39	-07	30	*	21	13	11
(12) PMA S_1	08	-08	04	11	-39	-42	-20	02	-27	-19	21	*	11	46
(13) PMA S_2	-07	12	-29	-23	21	-05	25	00	-08	14	13	11	*	-07
(14) Rod and Frame Test	00	-14	-19	-09	-22	-07	03	-01	-20	18	11	46	-07	*

Correlations of .37 are significantly different from zero at $p < .01$; correlations of .29 at $p < .05$.

ACKNOWLEDGMENTS

The preparation of this paper, and most of the results reported in it, was supported by grants 5-K3-MH 18,382, MH 07880, MH 07708, and HD 00843 of the National Institutes of Health and by grant GB 466 from the National Science Foundation.

The testing of high school students would not have been possible without the cooperation and encouragement of Richard VanHoose, Superintendent of Schools in Jefferson County, and Samuel V. Noe, Superintendent of Schools in Louisville and of the Very Rev. Msgr. Alfred Steinhauser, the Director of Catholic Schools for the Archdiocese of Louisville, Kentucky.

All blood group tests were performed at cost by Mrs. Jane Swanson of the Minneapolis War Memorial Blood Bank. I am indebted to Dr. Albin B. Matson for making this arrangement possible. Robert J. Long, M.D., and Robert G. Howard, M.D., collected the blood samples, while they were going to medical school.

The following individuals assisted in various ways in the collecting of data: Mrs. Patricia McGinty, Mrs. Elsie Long, Mrs. Thelma Gliessner, Mrs. Nancy Singleton, Robert G. Howard, Marlee Thompson, Mrs. Charlotte Russell, Maurice LeCroy, Philip Buckman, and Steve Potts.

Some of the statistical analyses were performed by Dr. James Lingoes, Dr. John Loehlin, and Dr. David Saunders.

Most important of all was the willingness of the hundreds of twins who participated in these experiments.

BIBLIOGRAPHY

BARROWS, S. L.
1945. The inheritance of the ability to taste brucine. Unpublished master's thesis, Stanford University.

BARTLETT, M. S.
1950. Tests of significance in factor analysis. *Brit. J. Stat. Psychol.*, 3:77–85.

BLEWETT, D. B.
1954. An experimental study of the inheritance of intelligence. *J. Ment. Sci.*, 100: 922–933.

CARTER, H. D.
1932. Twin similarities in occupational interests. *J. Educ. Psychol.*, 23:641–655.

CATTELL, R. B.
1953. Research designs in psychological genetics with special reference to the multiple variance method. *Amer. J. Hum. Genet.*, 5:76–91.

CATTELL, R. B., H. BELOFF, and R. W. COAN
1958. *Handbook for the IPAT high school personality questionnaire.* Champaign, Ill.: Institute of Personality Testing.

CATTELL, R. B., D. B. BLEWETT, and J. R. BELOFF
1955. The inheritance of personality. *Amer. J. Hum. Genet.*, 7:122–146.

CLARK, P. J.
1956. The heritability of certain anthropometric characters as ascertained from measurements of twins. *Amer. J. Hum. Genet.*, 8:49–54.

CONTERIO, F., and B. CHIARELLI
1962. Studio sulla ereditabilita di alcuni caratteri antropometrici misurati sui gemelli. *L'Atheneo Parmense*, 33 Suppl. 1.

DAHLBERG, G.
1926. *Twin births and twins from a hereditary point of view.* Stockholm: Tidens.

EYSENCK, H. S., and D. B. PRELL
1951. The inheritance of neuroticism, an experimental study. *J. Ment. Sci.*, 97:441–465.

FLEISHMAN, E. A.
1960. Psychomotor tests in drug research. In L. Uhr and J. G. Miller (Eds.), *Drugs and behavior.* New York: John Wiley.

FRUCHTER, B.
1950. Error scores as a measure of carefulness. *J. Educ. Psychol.*, 41:279–291.

FULLER, J. L., and W. R. THOMPSON
1960. *Behavior genetics.* New York: John Wiley.

GARDNER, R. W.
1964. The development of cognitive structures. In Constance Scheerer (Ed.), *Cognition: theory, research, promise.* New York: Harper and Row.

GOTTESMAN, I. I.
1963. Heritability of personality: a demonstration. *Psychol. Monogr.*, 77, No. 9, Whole Number 572.

GUTTMAN, RUTH, and L. GUTTMAN
1963. Cross-cultural stability of an intercorrelation pattern of abilities: a possible test for biological basis. *Hum. Biol.*, 35:53–60.

HARPER, J. C., C. C. ANDERSON, C. M. CHRISTENSEN, and S. M. HUNKA (Eds.)
1964. *The cognitive process, readings.* Englewood Cliffs, N.J.: Prentice-Hall.

HIRSCH, J.
1963. Behavior genetics and individuality understood. *Science*, 142:1436.

HOLT, SARAH B.
1961. Quantitative genetics of fingerprint patterns. *Brit. Med. Bull.*, 17:247–250.

HOROWITZ, S. L., R. H. OSBORNE, and FRANCES V. DEGEORGE
1958. Heredity factors in tooth dimension, a study of the anterior teeth of twins. *Angle Orthodontist*, 28:87–92.

HUNTER, W. S.
1959. The inheritance of mesiodistal tooth diameter in twins. Unpublished doctoral dissertation, University of Michigan.

JONES, L. V.
1949. A factor analysis of the Stanford-Binet at 4 age levels. *Psychometrika*, 14: 299–331.

KRYSHOVA, N. A., Z. V. BELIAEVA, A. F. DMITRIEVA, M. A. ZHILINSKAIE, and L. G. PERNOV
1963. Investigation of the higher nervous activity and of certain vegetative features in twins. *Soviet Psychol. Psychiat.*, 1:36–41.

LOEHLIN, J. C.
1965. A heredity-environment analysis of personality inventory data. In S. G. Vandenberg (Ed.), *Methods and goals in human behavior genetics.* New York: Academic Press.

LOUISVILLE CHAMBER OF COMMERCE
1962. *Market data by census tract, Louisville metropolitan area.* Louisville, Kentucky.

LUNDSTRÖM, A.
1948. *Tooth size and occlusion in twins.* New York: Karger.

1955. The significance of genetic and non-genetic factors in the profile of the facial skeleton. *Amer. J. Orthodontics*, 41:910–916.

LYSELL, L.
1955. *Plica palatinae transversae and papilla incisiva in man.* Uppsala: Almqvist and Wiksell.

MALAN, M.
1940. Zur Erblichkeit der Orientierungs-fähigkeit im Raum. *Zsch. Morphol. Anthrop.*, 39:1–23.

MEYERS, C. E., and H. F. DINGMAN
1960. The structure of abilities at the preschool ages: hypothesized domains. *Psychol. Bull.*, 57:514–532.

MYERS, ISABELLE BRIGGS
1962. *The Myers-Briggs type indicator.* Princeton, N.J.: Educational Testing Service.

NEWMAN, H. H., F. N. FREEMAN, and V. J. HOLZINGER
1937. *Twins, a study of heredity and environment.* Chicago: University of Chicago Press.

NICHOLS, R. C.
1965. The National Merit twin study. In S. G. Vandenberg (Ed.), *Methods and goals in human behavior genetics.* New York: Academic Press.

NOLAN, E. G.
1958. Uniqueness in monozygous twins. Unpublished doctoral thesis, Princeton University.

OSBORNE, R. H., and FRANCES V. DEGEORGE
1959. *Genetic basis of morphological variation.* Cambridge, Mass.: Harvard University Press.

PICKFORD, R. W.
1951. *Individual differences in color vision.* London: Routledge and Kegan Paul.

PONS, J.
1958. *Relaciones entre grupos sanguinos y lineas dermo-papilaris en negros de la Guinea española.* Madrid: Librería Científica Medinacal.

POST, R. H.
1962a. Population differences in red and green color vision deficiency: a review, and a query on selection relaxation. *Eugen. Quart.*, 9:131–146.

1962b. Population differences in vision acuity: a review with speculative notes on selective relaxation. *Eugen. Quart.*, 9:189–212.

REISS, A. J.
1961. *Occupation and social status.* New York: Free Press.

RIESMAN, D.
1950. *The lonely crowd, a study of the changing American character.* New Haven, Conn.: Yale University Press.

RITTER, H.
1958. Zur Morphologie und Genetik normaler mesodermaler Iristrukturen. *Zsch. Morph. Anthrop.*, 49:148–195.

SANK, DIANE

1963. Genetic aspects of early total deafness. In J. D. Rainer, K. Z. Altshuler, F. J. Kallman, and W. E. Deming (Eds.), *Family and mental health problems in a deaf population.* New York State Psychiatric Institute Columbia University.

SAVARA, B. S.

1963. Growth and development of the face. Progress Report to National Institutes of Dental Health, University of Oregon Dental School, Portland.

1965. Applications of photogrammetry for quantitative study of tooth and face morphology. *Amer. J. Phys. Anthrop.,* 23:427–434.

SCHEERER, CONSTANCE (Ed.)

1964. *Cognition: theory, research, promise.* New York: Harper and Row.

SMITH, G.

1949. *Psychological studies in twin differences.* Lund: Gleerup.

1953. Twin differences with reference to the Müller-Lyer illusion. *Lunds Universitet Arsskrift N.F. Aud. 1,* 50:1–27.

SNIJDERS, J. TH., and N. SNIJDERS-OOMEN

1958. Non-verbal intelligence test for normal and deaf children. Groningen, The Netherlands: J. B. Wolters.

SNYDER, L., and D. F. DAVIDSON

1937. Studies in human inheritance. XVIII: The inheritance of taste deficiency to diphenylguanidine. *Eugen. News,* 22:1–2.

SPUHLER, J. N.

1951. Genetics of three normal morphological variations: patterns of superficial veins of the anterior thorax, peroneus tertius muscle, and number of vallate papillae. *Cold Spr. Harb. Symp. Quant. Biol.,* 15:175–189.

STRONG, E. K.

1955. *Vocational interests 18 years after college.* Minneapolis: University of Minnesota Press.

THURSTONE, L. L., and THELMA G. THURSTONE

1941. *The primary mental abilities tests.* Chicago: Science Research Associates.

THURSTONE, THELMA G., L. L. THURSTONE, and H. H. STRANDSKOV

1955. *A psychological study of twins.* Chapel Hill, N.C.: Psychometric Laboratory, University of North Carolina, Report No. 4.

VANDENBERG, S. G.

1962a. The hereditary abilities study: hereditary components in a psychological test battery. *Amer. J. Hum. Genet.,* 14:220–237.

1962b. How "stable" are hereditary estimates? A comparison of heritability estimates from six anthropometric studies. *Amer. J. Phys. Anthrop.,* 20:331–338.

1964. The developmental study of twins. *Amer. Psychol.,* 19:537.

1965. Multivariate analysis of twin difference. In S. G. Vandenberg (Ed.), *Methods and goals in human behavior genetics.* New York: Academic Press.

VANDENBERG, S. G., P. J. CLARK, and INA SAMUELS

1965. Psychophysiological reactions of twins; heritability estimates of galvanic skin resistance, heart beat and breathing rates. *Eugen. Quart.,* 12:7–10.

VANDENBERG, S. G., and LILLIAN KELLY

1964. Hereditary components in vocational preferences. *Acta Genet. Med. Gemell.,* 13:266–277.

VANDENBERG, S. G., and H. H. STRANDSKOV
  1964. A comparison of identical and fraternal twins on some anthropometric measures. *Hum. Biol.*, 36:45–52.

VOGEL, F., and H. E. REISER
  1960. Zwillingsuntersuchungen über die Erblichkeit einiger Zahnmasse. *Anthrop. Anz.*, 24:231–241.

VOGEL, F., and G. G. WENDT
  1956. Zwillingsuntersuchung über die Erblishkeit einiger anthropologischer Masse und Konstitutions-indices. *Zsch. menschl. Vererb. u. Konstitutionslehre*, 33:425–446.

VON BRACKEN, H.
  1939. Wahrnehmungstäuschungen und scheinbare Nachbildgrösze bei Zwillingen. *Arch. ges. Psychol.*, 103:203–230.

VON VERSCHUER, O.
  1954. *Wirksame Faktoren im Leben des Menschen.* Wiesbaden: Franz. Steiner.

WITKIN, H. A., H. B. LEWIS, M. HERTZMAN, K. MACHONER, P. B. MEISNER, and S. WAPNER
  1954. *Personality through perception.* New York: Harper.

RELATION OF BEHAVIORAL, GENETIC, AND NEUROENDOCRINE FACTORS TO THYROID FUNCTION[1]

DAVID A. HAMBURG AND DONALD T. LUNDE

CHANGES IN THYROID FUNCTION ASSOCIATED WITH EMOTIONAL DISTRESS IN MAN

WHILE THE SIGNIFICANT RESPONSE of the adrenal cortex to stress in man has been an area of intensive investigation yielding generally consistent results in recent years (Hamburg, 1966), the thyroid's capacity to respond to psychological stress has remained in considerable doubt. For many years, clinicians in psychiatry, internal medicine, and pediatrics have been impressed by the occasional occurrence of hyperthyroidism (thyrotoxicosis) immediately following an extremely disturbing experience; but little has been learned of mechanisms through which such events are brought about. More generally, the thyroid response to psychological stress has appeared less consistent and of smaller magnitude than the adrenocortical and adrenomedullary responses. It has been difficult to clarify the conditions under which a significant change in thyroid function occurs in response to stress. In the present paper, we wish to examine some of the possible reasons for this difficulty, and to suggest the relevance of genetic factors to this problem.

We hope to call attention to several new lines of evidence that have potential importance for the analysis of human stress responses. These lines of evidence include: (1) a growing body of experimental literature documenting CNS influences on thyroid function; (2) a substantial body of research in the framework of human biochemical genetics documenting genetic sources of variation in secretion and transport of thyroid hormones, including several abnormalities of clear clinical significance; (3) a cumulative record of clinical and experimental evidence indicating powerful effects of thyroid hormone on brain and behavior, especially in a developmental perspective; and (4) a tendency in stress research and psychosomatic medicine to neglect a variety of provocative observations bearing on the possible significance of undersecretion of thyroid hormones in some circumstances of emotional distress. Altogether, by examining thyroid function as it relates to brain and behavior, we hope to indicate the promise of an integrated behavior-endocrine-genetic approach to stress problems.

Some thirty years ago Mittlemann (1933) studied sixty cases of hyperthyroidism and noted that a definite emotionally traumatic precipitating event had occurred in 49 of the cases, usually one or two weeks prior to the onset of symp-

1. This work was supported by N.I.H. Grant MH-10976.

toms. Lidz (1949) studied fifteen cases of hyperthyroidism and found that an event which terminated or threatened to terminate an essential interpersonal relationship for the patient had occurred immediately prior to the onset of symptoms in fourteen of the fifteen cases. Kleinschmidt *et al.* (1956) found that a traumatic event was chronologically related to the onset of thyrotoxicosis in 85 per cent of a large series of cases. Similar findings have also been reported by Nemeth and Ruttkay-Nedecky (1958), although one group of Russian workers has reported a much lower incidence of psychic trauma preceding diffuse toxic goiter (Baranov *et al.*, 1961). Hinman *et al.* (1957) emphasized traumatic experiences of loss or separation in the etiology of hyperthyroidism, as have other clinical investigators. They also noted an increased incidence of such experiences (for example, the death of a sibling) in the early years of the patients' lives. They suggested that hyperthyroid patients regress to an earlier stage of fixation when faced with a separation trauma, and that the hyperactivity of the thyroid in childhood is reactivated. According to their hypothesis, this heightened activity is absorbed by the physiological processes of maturation in childhood, but this outlet is not available in the adult, and hyperthyroidism occurs.

In addition to these anamnestic studies, direct observations of thyroid function have been made under conditions of naturally occurring and experimental stress. A study of protein-bound iodine (PBI) blood levels, as an index of thyroid function, in thirty psychiatric patients immediately upon admission to a general hospital in relatively intense emotional distress, showed a tendency toward PBI elevations (Board, Persky, and Hamburg, 1956). In general, these patients had been experiencing distress for a considerable period, typically several weeks. The patients' PBI mean level was clearly higher than that of normal control subjects, but the difference was proportionately less than that observed for corticosteroids. Analysis of hormone levels of patients in psychiatric subgroups indicated that unusually high hormone levels were associated with extremely intense distress and with personality disintegration. When patients were retested about two weeks later, following intensive treatment, PBI levels were similar to those of the initial day. When this follow-up group was divided into two subgroups, consisting of those who did and did not receive electroconvulsive therapy (ECT), they differed in mean levels of PBI (and corticosteroids). The more disturbed patients who had received ECT had relatively high PBI levels, while the less disturbed patients who had not received ECT had lower PBI levels. The observations of this study were generally consistent with the concept of modest PBI elevations under conditions of relatively intense emotional distress of a few weeks duration.

In an extension of this work, a series of 33 consecutive depressed patients were studied immediately upon admission to the psychiatric section of a general hospital (Board, Wadeson, and Persky, 1957). Protein-bound iodine levels were again elevated in comparison with normal control subjects, but not to a statistically significant extent. On a later occasion of testing, the PBI levels were lower, concomitantly with a fall in ratings of depressive intensity.

A study of psychologically normal subjects under stress yielded somewhat ambiguous results (Volpe *et al.*, 1960). Nine of eleven medical students who were observed during the academic year were found to have their highest PBI levels during the examination period, although the elevations were regarded as slight by the investigators. Football players, surgical patients, and patients with recent myocardial infarcts were also studied, but correlations of PBI changes with stressful experiences were not detected in these subjects.

The use of stressful interviews based on information from the patient's history has indicated that this sort of experience may affect thyroid function. Hetzel and his coworkers (1952, 1956) demonstrated rises in PBI shortly after such interviews in both euthyroid and hyperthyroid patients. However, they did not see this effect in subjects who were initially hypothyroid. Using a similar procedure, another group failed to find any changes in PBI after stressful interviews (Dongier *et al.*, 1956).

An interesting study by Franz Alexander and his colleagues (1961) used a stress situation which was more standardized than the interviews, that is, an exciting movie with a plot centered around the fear of death. Hyperthyroid and normal subjects were studied. An increase in PBI[131] was seen in the hyperthyroid patients immediately after the movie. This effect was particularly striking in the case of seven subjects who were acute hyperthyroids. The control subjects showed an initial drop in PBI[131], but a rise one hour later to values that exceeded their previewing levels.

In addition to the studies of stress and thyroid function in humans, there are a few animal studies that are relevant here. Kracht (1954) reported the occurrence of a so-called fright thyrotoxicosis in captured wild rabbits, although another group of investigators was unable to confirm this finding (Brown-Grant, Harris, and Reichlin, 1954). The same group reported that stress in the form of electric current, restraint, changes in environmental lighting, or various physical traumata produced an *inhibition* of thyroid function as indicated by a decrease in the release of I[131] labelled hormone from the thyroid gland of rabbits. Neither denervation of the thyroid nor adrenalectomy appeared to prevent this inhibitory effect. An experimental model for chronic stimulation of thyroid hormone secretion might be developed in analogy to the situation sometimes seen in hyperthyroid patients that Kriss (personal communication) described as "psychological entrapment." He described this as an unremitting stress of relatively long duration which is viewed as catastrophic by the patient and which seems to offer no escape; for example, a spouse with an incurable debilitating disease, or a husband who becomes psychotic and is physically threatening to his wife, who in turn becomes hyperthyroid while living a captive existence, afraid to consult medical or legal authorities.

Reiss (1956) reported a decrease in I[131] uptake of 70 per cent in rats that have been swimming in cold water for fifteen minutes. He also noted a reduction in thyroid activity after giving a single electroshock to the animals. Interestingly enough, he noted these results in both intact and hypophysectomized animals, and concluded that this acute stress effect was related to the release of vasoconstrictor substances. Twenty-four hours after swimming or electroshock, however, the I[131] uptake was significantly increased.

Changes in Thyroid Function Associated with Personality Characteristics and Psychiatric Disorders in Man

Numerous studies have emphasized the finding of a common personality pattern in patients who become hyperthyroid (Mittlemann, 1933; Lidz, 1949; Ham, Alexander, and Carmichael, 1951; Bennett and Cambor, 1961; Lubart, 1964). Most of these studies were done within a psychoanalytic orientation, and emphasized the element of unresolved dependency needs in these patients. The frustration of dependent longings in childhood and threats to security have been viewed as leading to premature attempts at independence in these patients. The persistence of unsatisfied dependency motivation makes them particularly vulnerable to separation experiences, and such events are thought to precipitate symptoms of hyperthyroidism.

These exploratory, clinical investigations are retrospective and subject to observer bias of unknown magnitude. However, a later prospective study based on these earlier observations (Dongier et al., 1956) found some predictive validity in the psychodynamic formulation of the hyperthyroid personality. They found that 12 of 44 neurotic patients conformed to the personality configuration described by the writers previously mentioned, and of these 12, 9 had high thyroid secretion rates as indicated by I[131] disappearance curves. Only 7 of the other 32 subjects had high secretion rates. The association between personality configuration and thyroid secretion was found to be statistically significant. However, a more recent and quite careful study by Hermann and Quarton (1965) has not provided support for the by now classical hypotheses regarding specific personality and family patterns in the etiology of hyperthyroidism.

Also, efforts to identify a common personality type among hyperthyroid patients by the use of psychological tests have so far not been successful. Kleinschmidt et al. (1956) was unable to identify a specific personality structure in his series of 84 cases of thyrotoxicosis; but he pointed out that experiences of deprivation and frustrated dependency needs, including actual abandonment in childhood, were common features in the histories of these patients. Jones (1959) reported similar results.

Psychological tests have been shown to discriminate between acute thyrotoxic patients and normals or between various groups of psychiatric patients (Robbins and Vinson, 1960). Prior to thyroid treatment, the only group the thyrotoxic patients resembled in test responses was a group of patients with organic brain disease. However, after treatment the thyrotoxic subjects could no longer be discriminated from the normals.

In addition to personality types, some attempts have been made to correlate thyroid function with various psychiatric diagnoses. The data here are equivocal, and may indicate that current psychiatric diagnostic categories simply do not relate in any consistent way to thyroid function; there is perhaps no reason to expect that they should.

There has been particular interest in the thyroid function of schizophrenics. While it has been said that the basal metabolic rate is abnormally low in about 50 per cent of schizophrenic patients, one study reported that schizophrenics have a significantly higher iodine uptake than controls (Bowman et al., 1950). Certain other variables must be taken into account here, such as the diet of chronic mental patients. One study of psychotic patients with high iodine uptakes showed that this finding could be negated by the introduction of iodized salt into the diet (Kelsey and Gullock, 1957). It has also been pointed out that other factors besides the output of thyroid hormone may be relevant to the level of oxygen consumption in these patients (Brody and Man, 1950).

It has been reported that rather severely depressed patients in a general hospital, particularly those with a high degree of psychomotor retardation, tend to have elevated PBI levels (Board, Wadeson, and Persky, 1957). Further studies on various types of depressed patients would be in order, as well as comparisons of depressed patients with other severe disorders.

A study of eleven patients selected for intense free-floating anxiety and aggression showed PBI levels to be within the normal range (Kleinschmidt et al., 1956); but in a much larger series of cases studied by Reiss et al. (1953), a tendency toward thyroid hyperfunction was found in anxious female patients, whereas a tendency toward hypofunction was found in anxious male patients. In all, Reiss studied 1,000 psychiatric patients with various diagnoses, and found that thyroid function was not normal (that is; it was either high or low) in 20 per cent of the cases, as indicated by radioactive iodine uptakes. An interesting finding in this study concerns 76 of the patients who were either hyper- or hypothyroid prior to treatment (which consisted of electroshock therapy, insulin coma, CO_2, leucotomy, psychotherapy, or various combinations). There was a statistically significant association between psychiatric improvement with treatment and normalization of thyroid function in this group.

Most of the literature reviewed in this section suggests that personality characteristics or psychiatric disorders are associated in a causal way with abnormalities of thyroid function. Studies of larger numbers of patients using more exacting psychological methods and a greater variety of measures of thyroid function are needed to clarify this interesting possibility. One might also consider the opposite side of the coin: do changes in thyroid hormone levels, either as the result of disease or manipulation by the physician, elicit subsequent changes in the behavior of patients? This problem will be considered in a later section.

Biosynthesis, Secretion, and Catabolism of Thyroid Hormones

To facilitate understanding of the recent research literature, we will present in the next two sections a brief review of thyroid function and its assessment. Thyroid hormone synthesis begins with concentration of iodide in the gland by means of the thyroid iodide "trap" or "pump," capable of achieving concentrations of iodide ions in the thyroid gland that are 25 to 500 times the concentration found in plasma (Williams and Bakke, 1962). The thyroid iodide pump is apparently under TSH control (Pitt-Rivers, 1960). After being trapped in the gland, the iodide is oxidized to iodine by means of a peroxidase enzyme. Iodination of tyrosine then forms mono-iodotyrosine (MIT) and further iodination of this compound produces di-iodotyrosine (DIT). Two molecules of DIT can then be coupled to form thyroxin (T4) or a molecule of MIT and a molecule of DIT can be condensed to form tri-iodothyronine (T3) (Tepperman, 1962).

The compounds MIT, DIT, T3, and T4 are not free in the gland in any appreciable amount, but are bound to a large protein molecule by means of a peptide linkage with tyrosine. This protein, thyroglobulin, has a molecular weight of about 680,000 and prevents the iodinated compounds from entering the capillaries of the gland. A protease enzyme, stimulated by TSH, causes release of the iodinated compounds from thyroglobulin, and thyroxin, along with smaller amounts of T3, is then released into the bloodstream. MIT and DIT are deiodinated while still in the thyroid gland by a deiodinase enzyme, and in this way iodide and tyrosine are salvaged and can once again enter the cycle (Williams and Bakke, 1962; Tepperman, 1962).

In the circulating blood, some thyroxin is in the "free" form, but much of it is strongly bound to one of several thyroxin-binding proteins (TBP), and is thus inactive (Imarisio and Greco, 1964). Free thyroxin can enter cells where it stimulates oxygen consumption and energy release; it is then degraded or inactivated by any of several transformations, including conjugation with glucuronide or sulfate, decarboxylation, deamination, deiodination, oxidation of the phenyl group, or uncoupling of the linkage between the two rings of the molecule (Williams and Bakke, 1962). The water-soluble glucuronide conjugate is excreted in the bile, but a portion of the iodine thus excreted may be reabsorbed from the intestine after hydrolysis of the conjugate by intestinal bacterial β-glucuronidase (Williams and Bakke, 1962).

The average biological half-life of thyroxin in normal individuals has been found to be about 6.7 days (Sterling and Chodos, 1956).

Assessment of Thyroid Function

In order to evaluate research on thyroid problems, it is essential to have some understanding of the myriad diagnostic tests that may be used to measure thyroid function. These tests do not all measure the same processes, and each is subject to different sources of variation and error.

One of the most common laboratory tests is the chemical PBI (protein-bound iodine) (Sunderman, 1963). This test is based on the fact that most of the thyroxin in the serum is bound to proteins and is precipitated with them when exposed to any of several denaturing agents. The amount of iodine in this precipitate is measured and is proportional to the amount of hormone present, but the test will be invalid if the subject has taken iodine-containing drugs recently. Certain iodine-containing X-ray contrast media may cause artefactual depression by inhibiting the color reaction used in the assay. Since the iodine measured is related to the amount of protein present, conditions associated with an increase in thyroxin-binding protein (pregnancy, hepatitis, familial trait) will lead to a high PBI value, whereas conditions associated with TBP deficiency (nephrosis, chronic debilitating disease) will result in a low PBI. In spite of its limitations, the PBI has been said to be "perhaps the most accurate index of thyroid hormone available" (De Groot, 1960), although it should be used in conjunction with other tests of thyroid function, particularly in doubtful cases of hyperthyroidism (Radcliff et al., 1964).

The PBI[131] is a similar test, based on prior administration of I[131] (De Groot, 1960). It can be artefactually low due to dilution of the I[131] by large amounts of iodine in the plasma, resulting from administration of Lugol's solution, potassium iodide, or iodine-containing drugs, cough syrups, or tonics. Some of these sources of potential error are eliminated by the newer method of measuring "T_4 by column," which has recently been adopted by some commercial laboratories.

The BEI (butanol-extractable iodine) is a measure said to be "more specific diagnostically for patients who have taken inorganic iodine or have inborn errors of synthesis of metabolically active iodine compounds" (Man, 1963). This test eliminates an elevation that might be due to the presence of iodinated proteins as in Hashimoto's thyroiditis, thyroid carcinoma, or congenital defects in hormone synthesis, since it does not measure iodoproteins (De Groot, 1960).

Another standard test is the uptake of radioiodine by the thyroid (Burkle and Lund, 1963). The uptake of radioiodine over a 24-hour period (measured by a counter over the neck) is considered to be an index of the activity of the gland. Numerous drugs are known to diminish the I[131] uptake, however, including perchlorates, nitrates, thyroid compounds, estrogens, aspirin, penicillin, and vitamins. The 24-hour uptake may also be misleading in thyrotoxic patients where the uptake and secretion of iodine is so rapid that, by the end of 24 hours, a large part of the I[131] taken up may have already left the gland as labelled hormone (Goolden, 1960).

The per cent red cell uptake of I[131]-triiodothyronine (T_3) test is designed to measure the degree of saturation of TBP binding sites (Burkle and Lund, 1963). In hyperthyroidism there will be fewer binding sites available on plasma proteins due to the high concentration of hormone present; consequently a greater percentage of the labelled T_3 will be bound to red cells, when added to a sample of whole blood from such a patient. An unexplained finding is an unusually high red cell uptake in mongoloid children. This occurs in the absence of any signs of thyrotoxicosis and in the presence of normal thyroxin binding proteins and a normal turnover rate for thyroxin (Ferrier and Lemarchand-Berand, 1965). Conditions like nephrosis, liver disease, pregnancy, or ingestion of estrogens (for example, oral contraceptives) will alter the red cell uptake because of their effect on plasma protein levels.

A new test, which apparently is not affected by exogenous iodine, estrogens, or pregnancy, is the tyrosine tolerance test (Rivlin, Melmon, and Sjoerdsma, 1965). This test is based on the observation that plasma levels of tyrosine are greatly elevated after oral loading in hyperthyroid patients, but are decreased in hypothyroid patients.

In some cases, it is important to know the amount of thyroxine-binding proteins present in the serum, and such determinations can be made by use of paper electrophoresis or other methods (Robbins, 1963).

Several tests that estimate the state of thyroid function indirectly are also used by some investigators; they should be used in conjunction with more direct methods that have already been described.

The basal metabolic rate (BMR) is a classic test which has become rather unpopular in recent years (Smetters, 1963). Many factors other than hyperthyroidism can produce an elevation of the BMR, including smoking, anxiety, lack of sleep, muscular activity, eating, dyspnea, certain neoplastic diseases, and fever. The test may be a useful adjunct in the diagnosis of hypothyroidism, however. An inverse relationship between serum cholesterol levels and thyroid activity exists, and attempts have been made to use serum cholesterol as an index of thyroid function (Peters and Man, 1950). However, cholesterol levels are also affected in nephrosis, diabetes, and liver disease, and may change in response to diet, drugs, and age. Although a low cholesterol level is not particularly useful in establishing the diagnosis of hyperthyroidism, an elevated serum cholesterol may be useful in establishing the diagnosis of hypothyroidism, particularly in children (Kriss, 1964).

One other indirect test which might be mentioned is the Achilles tendon reflex (Sherman *et al.*, 1963). This test involves measurement of the duration of the Achilles reflex by means of a photoelectric apparatus. The reflex is slowed in hypothyroidism and faster than normal in hyperthyroidism. Results of this test compare quite favorably with more conventional measures; and the test has the advantages of being simple to perform and unaffected by iodine-containing drugs or conditions like pregnancy.

CNS CONTROL OF THYROID[1]

With this background of information on thyroid function and its assessment, we can now proceed to a consideration of thyroid-CNS relations. Is there evidence that the central nervous system participates in the regulation of secretion of thyroid hormone? Is there evidence that the circulating thyroid hormone influences brain and behavior? To the extent that these questions can be answered affirmatively, there is encouragement to pursue the problem of thyroid responses in psychological stress—in spite of the equivocal results we have earlier described.

Since the thyroid is capable of responding to an external stimulus like cold, it seems likely that the nervous system must be involved in regulating the secretion

1. For a more detailed coverage of this subject, one should consult the excellent review by S. A. D'Angelo in A. V. Nalbandov (Ed.), *Advances in Neuroendocrinology* (Urbana: University of Illinois Press, 1963), Chap. 6.

of this gland, yet only in recent years has a substantial body of research in neuroendocrinology been developed. Much remains to be done in clarifying the brain's influences on thyroid function.

It is well known that thyroid activity is dependent upon secretions from the pituitary; and several types of experiments have demonstrated that the pituitary, in turn, is subject to control by the central nervous system via the hypothalamus. Before discussing these experiments, we should point out that there are two different modes of control of the pituitary over the thyroid (Reichlin, 1963b). One of these is a simple feedback mechanism which involves sensitivity of the anterior pituitary to levels of circulating thyroid hormones. The other mechanism is a neural one, and involves "setting" the level at which the pituitary-thyroid system will respond, much as one would set a thermostat. Indeed, this mode of control has been referred to as the pituitary "thyrostat" (Reichlin, 1963b).

One must also keep in mind that stimulation of the thyroid by TSH (thyroid-stimulating hormone, thyrotropin) involves two aspects: the release of stored hormones into the blood, producing a more immediate and short-term effect; and an increased rate of synthesis of thyroid hormones, a more prolonged, long-term effect (Purves, 1964).

Experiments demonstrating the importance of CNS-pituitary connections have followed several patterns. The simplest, perhaps, has involved sectioning the pituitary stalk and observing changes in thyroid function. Such studies have shown that the pituitary feedback system continues to operate at a baseline level (Uotila, 1939, 1940), with some drop in I^{131} uptake and I^{131} release rate reported (in rabbits), though it is not as great a change as that seen in hypophysectomized animals (Brown-Grant, Harris, and Reichlin, 1957). Uotila (1940) found that animals in which the pituitary stalk had been sectioned no longer showed a thyroid response to cold.

Another type of experiment has involved transplantation of the pituitary to some site remote from the hypothalamus, for example, the kidney or the anterior chamber of the eye. One such experiment (Von Euler and Holmgren, 1956) involved transplants of the pituitary to the eye in rabbits. Again, it was found that a certain degree of thyroid secretion was maintained and that the pituitary feedback system (tested by observing a decrease in thyroid secretion when exogenous thyroxin is administered) was still intact. But in these same animals, exposure to cold, which would be expected to change the "thyrostat" setting for a higher level of secretion to be maintained, no longer had any effect.

The standard techniques of neural lesions and stimulation have also been employed in this field. Harris and Woods (1958) showed that electrical stimulation of the anterior hypothalamus of rabbits produces an increase in the rate of release of I^{131} from the thyroid gland and an increase in PBI^{131}. Stimulation of the posterior hypothalamus does not produce these effects.

It has also been shown that electrical stimulation of the anterior hypothalamus or the rostral portion of the median eminence in female rats produces an increase in TSH levels (D'Angelo and Snyder, 1963). This increase in TSH could be prevented by lesions in these same areas.

Numerous studies have shown that hypothalamic lesions can produce a decrease in thyroid function, and the consensus is that the area most specific for altering thyroid function, as in the stimulation studies, is in the rostral part of the hypothalamus (D'Angelo, 1963). A recent report from Hungary confirmed the specificity of this anterior area for thyroid control, pointing out that lesions farther down in the region of the posterior median eminence and the hypophyseal stalk may also affect the thyroid, but other endocrine glands are then involved as well (Halasz et al., 1963). It appears that the latter region might well constitute a "final common pathway" for neural influences of the hypothalamus on the various endocrine glands, although there are still too many equivocal reports to consider the matter settled (D'Angelo et al., 1964).

Another sort of experiment that implicates the hypothalamus in control of thyroid function has involved localizing cooling of certain brain areas. Andersson et al. (1962) showed that cooling the pre-optic area of unanesthetized goats by means of permanently implanted thermodes produces an increase in PBI, which reaches a peak after about six hours.

The mechanism by which the hypothalamus influences the secretion of TSH by the pituitary is being investigated in several laboratories. The most widely accepted view at present is that certain cells of the hypothalamus are capable of releasing neurohumoral substances (probably polypeptides) that reach the anterior pituitary via the hypothalamo-hypophysial portal vessel system and are capable of causing the release of the various tropic hormones (Harris, 1962).

Shibusawa (1956) reported isolation of a thyrotropin-releasing factor (TRF) from dogs over ten years ago, although his results have been questioned (D'Angelo, 1963; Reichlin et al., 1963). More recently, others have reported the isolation of a TRF from bovine (Schreiber et al., 1962) and sheep hypothalamus (Guillemin et al., 1962). Guillemin (1963) reported that this substance is capable of causing the release of TSH from anterior pituitary tissue in amounts that are a function of the dose of TRF used.

It has been suggested that TRF may serve to stimulate pituitary TSH secretion, whereas circulating thyroid hormones act to inhibit TSH secretion (Reichlin, 1963a), with the actual output being a resultant of these two influences. One worker (Thomas, 1964) published results suggesting that both inhibition and increase of TSH secretion can be elicited by hypothalamic stimulation, depending on the location of the electrodes. Up to now, most workers have centered their attention on CNS influences that might increase thyroid activity. In the main, their view is that the hypothalamus is primarily concerned with increasing thyroid secretion, and that the center for such activity is in the supra-optic region, perhaps in the paraventricular nuclei (de Jong and Moll, 1965). It will be interesting to see whether future research will establish a neural substrate for inhibition of thyroid secretion, as has proven to be the case in regard to CNS influences on the adrenal cortex.

Mention should be made of the long-acting thyroid stimulator (LATS) that is found in the plasma of some hyperthyroid patients, the origin of which has been unknown. Recent work by Kriss *et al.* (1964) indicates that this substance may be an antibody to some thyroid constituent which is released as a result of damage to the gland in the course of treatment for hyperthyroidism. This work is particularly interesting because it points toward an antibody with potent hormonal properties, and one that probably has clinical significance.

Two other substances reported to affect thyroid function are melatonin (perhaps inhibitory) and vasopressin (said to stimulate the gland directly) (Baschieri *et al.*, 1963; Garcia, Harris, and Schindler, 1964). Further investigation should indicate whether these compounds are of any physiological significance in thyroid regulation. Areas of the brain other than the hypothalamus are probably capable of influencing thyroid function. Mess (1958, 1959) reported that lesions of the habenular nuclei in rats prevent the goitrogenic response to thiouracil and disrupt the normal pituitary-thyroid feedback system, although Greer (1963) could not confirm these results. It has also been reported that stimulation of the hippocampus in dogs will produce a six-fold increase in PBI within two hours (Shizume *et al.*, 1962). Another group has reported that lesions to the globus pallidus or septal areas stimulate release of TSH in rats, suggesting that these areas might normally have an inhibitory function (Lupulescu *et al.*, 1962).

Data in this area are not entirely consistent at the present time. Part of the problem appears to be that different authors use different assays for thyroid function and then make general statements about the effect of a particular lesion or stimulation on "thyroid function" in general. It seems important to distinguish between thyroid function in terms of the servomechanism that releases TSH from the anterior pituitary in response to a drop in circulating thyroxin levels and thyroid function in terms of the more complicated neural mechanisms of TSH control that involve response of the organism to external events. Thus, it has been shown that removal of the entire forebrain in rats, leaving only a "hypothalamic island," does not impair the pituitary response to low circulating thyroxin levels produced by propylthiouracil (Matsuda, Greer, and Duyck, 1963). However, observations on changes in thyroid secretion in response to psychological stress (described earlier) suggest that higher centers in the brain are capable of producing effects, possibly by altering the setting of the "thyrostat."

Effects of Thyroid Hormones (or Abnormalities in Thyroid Function) on Behavior in Man

Certain behavioral signs are often associated with thyroid dysfunction clinically (Means, 1948; Eayrs, 1960). The hyperthyroid patient is characterized by hyperexcitability, irritability, restlessness, increased sexual motivation, exaggerated responses to environmental stimuli, emotional instability, and, ultimately, psychosis. The hypothyroid patient is characterized by listlessness, lack of energy, slowness of speech, reduced sensory capacity, impairment of memory, decreased sexual motivation, and somnolence.

Very little work has been done in humans to correlate hormone levels with objective behavioral performance, although this sort of experiment has been done with various laboratory animals. One study of hyperthyroid patients did show an improvement after therapy on tests of auditory reaction time and time estimation (Stern, 1956), and similar results have been reported by Artunkal and Togrol (1964). The effects of administered thyroid hormones on avoidance conditioning in various learning experiments have been relatively small (Eayrs, 1960). However, when thyroid deficiency is produced in young rats during the course of cerebral development, the effects are quite marked. These cretinoid animals quickly become habituated to novel situations. They tend to persist in the performance of an established habit when faced with environmental changes to which a normal animal will react, and make more errors and spend less time in investigation as a result of an error than normal animals do in a maze-learning situation (Eayrs and Lishman, 1955). These effects are probably related to a defect in cerebral development which will be described in the next section.

Certain behavioral effects of thyroid hormones in adults are presumably related to physiological changes in the central nervous system rather than anatomical defects. In addition to the psychosis associated with severe, untreated thyrotoxicosis, there have been numerous reports of psychoses occurring subsequent to thyroidectomy (Carpelan, 1953; Gregory, 1956; Bonetti, 1958; Jones, 1959). In some of these cases there was evidence of psychiatric disorder prior to the thyroid disease that occasioned the thyroidectomy; but in other cases, the patient had been considered psychologically normal prior to surgery. The "post-thyroidectomy psychosis" has been described as a combination of perplexity, confusion, labile mood swings, and shifting, loose content disturbances (Brodeman and Whitman, 1952). Treatment with thyroid hormones has proved successful in some of these cases, indicating the possibility that the symptoms are due to thyroid deficiency. Indeed, there are several reports of psychotic behavior among myxedematous patients (Asher, 1949; Turmel et al., 1956; Rullo and Allan, 1958), and most of these have responded well to thyroid therapy.

The hypometabolic state of some schizophrenic patients has already been mentioned, and successful treatment of schizophrenic and depressive disorders with thyroid hormones (including one series of 29 post-partum depressions) has been reported (Danziger and Kindwall, 1948; Danziger, 1958; Feldmeyer-Reiss, 1958; Hamilton, 1962). However, in most of these cases the patients were receiving other forms of therapy as well, so one cannot attribute the results to thyroid alone. Other writers have reported only temporary behavioral changes in schizophrenic patients receiving thyroid, consisting of either decreased depression and withdrawal, increased restlessness and tension (Tolan and Dillon, 1960), or no change at all (Strisower et al., 1958; Simpson et al., 1964).

Present evidence certainly does not support any general relation between thyroid function and schizophrenia. In a Russian study, patients with manic syndromes were treated by thyroidectomy, whether or not they had indications of hyperthyroidism. In the series of patients reported, there were no relapses following this treatment (Parhon-Stefanescon and Vrejon, 1957). R. Gjessing (1938) reported cases of successful treatment of catatonic schizophrenia with thyroid hormones, but a more recent study of thyroid function in catatonic patients (L. R. Gjessing, 1964) was unable to find any abnormality that would help explain the pathogenesis of the disease or a mechanism for the reported beneficial effects of thyroid treatment.

Although it is well established that changes in behavior accompany the conditions of hypo- and hyperthyroidism, there has not been enough investigation beyond the clinical observations which have been described to indicate a mechanism for these changes, although they may be related to oxygen consumption within the central nervous system or interaction with adrenal medullary hormones, which will be discussed later.

Many of the above observations are complicated by the fact that thyroid hormones were not the only variable being manipulated. There is a need for research on the effects of thyroid hormones alone on normals as well as psychiatric patients in order to clarify what specific behavioral effects these compounds have in man. Concomitant study of the metabolic changes occurring in the subjects might provide further clues as to the mechanism of the behavioral changes.

EFFECTS OF THYROID HORMONES (OR ABNORMALITIES IN THYROID FUNCTION) ON BRAIN DEVELOPMENT AND FUNCTION IN MAN AND OTHER MAMMALS

The ability of various areas of the brain, particularly the hypothalamus, to concentrate radioactive labeled thyroid hormones has been demonstrated in the rabbit, the guinea pig, and the rat (Ford and Gross, 1958a, 1958b; Ford et al., 1959). It has also been reported that the uptake of thyroid hormones by cerebral gray matter and cerebellum is greater in female rats than in males (Ford et al., 1964). (This is one of several sex differences in thyroid function and disease that are known, and it would seem important to bear these differences in mind when working in this field, rather than grouping male and female subjects together, as is often done.)

The importance of thyroid hormones for normal development and function of the brain is well illustrated in the condition of cretinism. Among other things, the cretin often has a low IQ, but this and the other defects of this condition can be reversed to some extent by early and adequate replacement treatment with thyroid hormones (Money, 1956; Eayrs, 1960). The mechanism by which thyroid deficiency retards cerebral maturation is not known. Substances like methylandrostenediol and growth hormone, which stimulate protein anabolism and restore the growth of thyroidectomized animals, do not prevent or repair the CNS developmental defects (Eayrs, 1960). On the other hand, it was shown recently that L-thyroxin inhibits amino acid incorporation into protein in adult rat brain homogenates, but in infant brain preparations, thyroxin stimulates amino acid incorporation into protein when β-hydroxybutyrate is used as the substrate (Gelber et al., 1964). Another recent study showed that while neuronal maturation and myelinization is inhibited by low temperatures *in vitro*, this effect can be overcome by the addition of thyroid hormone, suggesting that thyroid may influence rate-controlled processes in a manner akin to temperature (Hamburgh and Bunge, 1964).

An alteration of brain function in cretins is reflected by abnormalities in the EEG, which have been described as a slowed basic frequency and flat featureless records (Beierwaltes *et al.*, 1959). The EEG tends to become normal if the patient with cretinism or infantile myxedema receives early and adequate thyroid treatment (Chaptl *et al.*, 1956; Di Gruttola, 1958).

Eayrs (1960) suggested a possible mechanism for these EEG abnormalities based on the anatomical finding that in infantile thyroid deficiency one finds a hypoplastic neuropil; in other words, there is a significant impairment of the development of nerve cell processes. The reduced probability of interaction between neurones associated with hypoplasia of the neuropil may be responsible for the changes that occur both in the electrical activity of the brain and the behavior of the cretin.

Hyperthyroidism occurs in childhood but there are apparently no striking effects on development in most cases. Management, however, is difficult with a high incidence of recurrence. There is evidence that there is a genetic factor operating in this somewhat unusual condition (Saxena, Crawford, and Talbot, 1964). In our experience, behavior disturbances similar to those of adult hyperthyroidism are impressive.

Changes in the EEG are also seen in the adult conditions of hypo- or hyperthyroidism. In general, there is a tendency toward slowing of the basic frequencies with thyroid deficiency and accelerated frequencies with thyroid excess (Vague *et al.*, 1958). Hypothyroid patients are found to have slowing or complete absence of alpha frequency (Nieman, 1959; Lansing and Trunnel, 1963). On the other hand, an increase in alpha frequency is reported in hyperthyroid patients, particularly in young women (Wilson and Johnson, 1964). Lansing and Trunnel (1963) pointed out that the alpha frequency does not normally fluctuate significantly, and that any change in alpha frequency of a half cycle or more within a few weeks represents a major deviation and may indicate a change in thyroid function.

In addition to EEG changes, certain neurological signs have been associated with hypothyroidism, including convulsions (Evans, 1960), a cerebellar syndrome (Jellinek and Kelly, 1960), and hemiplegia (Fotopulos, 1958). These neurological problems are usually reversible with thyroid treatment.

Although the development defects produced in the brain by congenital or infantile myxedema have been studied, less is known about the physiological defects that abnormalities of thyroid function produce in the brain. Some writers have stated that thyroid deficiency produces a decrease in cerebral oxygen and glucose consumption (Schienberg *et al.*, 1950), but others have not found this to be the case (Sensenbach *et al.*, 1954). Reiss (1956) was able to demonstrate an increase in cerebral oxygen consumption with thyroid administration in newborn rats, but he could not demonstrate this effect in intact adults. However, he did find that if the adult rats were pretreated with pentobarbital for several days, the oxygen consumption of the brain could be raised considerably by thyroxin treatment. The possible role of cerebral anoxia due to thyroid deficiency in psychiatric disorders has been suggested by some authors (Danziger and Kindwall, 1948), but the existence of such a relationship is doubtful.

There is some agreement as to the effects of thyroid on cerebral blood flow. Studies of patients with myxedema have shown that thyroid deficiency causes a decrease in cerebral blood flow and an increase in cerebrovascular resistance (Sensenbach et al., 1954). These conditions can be reversed with thyroid hormone.

Other possible ways in which thyroid hormones may bring about changes in brain function could have to do with the effects of thyroid on various enzyme systems or the interaction between thyroid and catechol amines. At present we must agree with Tepperman (1962), who said, "Since altered thyroid hormone availability has unmistakable effects on brain function *in situ* our inability to detect an obvious metabolic aberration . . .is an arresting finding." The effects of thyroid hormones on brain differentiation and metabolism represents an important research frontier. Such work should provide clarification of the interesting, unexplained findings in humans that have been described here.

GENETIC ABNORMALITIES IN THYROID FUNCTION

Numerous inborn errors in the thyroid pathway have been recognized in recent years (see Fig. 1). Stanbury (1963) described five of these defects, any one of which can produce familial goiter.

The consensus of genetic studies indicates that these defects are transmitted by autosomal recessive genes (Kitchin and Evans, 1960), although one recent paper (Leszynsky, 1964) described a family in which such a condition appeared to be transmitted by an incompletely dominant gene.

The five defects described in detail by Stanbury can be summarized as follows. First, there is a defect in the iodine trapping mechanism of the thyroid gland, such that the iodine concentration in the gland cannot be raised above that in the plasma. The abnormality here appears to be associated with an iodide active transport system that is common to the thyroid, salivary glands, and gastric mucosa. The parents of a child who was hypothyroid because of this defect had enlarged thyroids; and it is suggested that as heterozygotes for this condition, they had a relative deficiency of the iodide-concentrating ability that led to goiter formation.

It is of interest here that the conditions of endemic and sporadic goiter, rather than being solely related to a condition of iodine deficiency, are now thought to have a hereditary component, such that a subthreshold enzymatic deficit exists (quite possibly due to heterozygosity) and makes the individual much more sensitive to a slight iodine deficiency or ingestion of a goitrogen in the diet (De Visscher et al., 1964). Other defects may involve either of the proposed two steps involved in the organification of iodine. This includes an iodide peroxidase enzyme, to oxidize iodide to iodine, and an iodinase enzyme, to attach the iodine to tyrosine. It is possible that a deficiency of the latter enzyme is responsible for the condition called Pendred's syndrome, which is characterized by goiter, nerve deafness, and, in some cases, hypothyroidism. The eighth nerve damage is said to be related to hypothyroidism *in utero* during a critical period of development. Some individuals with this condition become euthyroid later in life, provided they have an adequate dietary iodine intake; but the damage to the acoustic nerve is irreversible. It is still not clear why all cretins are not deaf, however (Thould and Scowen, 1964). The frequency of the gene for Pendred's syndrome (a simple recessive trait), is estimated at between 1:150 and 1:500. It should be noted that defective organification of iodine is apparently the most common biochemical defect in goitrous cretins (Beierwaltes, 1964).

Another inborn error has to do with the mechanism for coupling the iodotyrosyl molecules to form the thyroid hormones. In such cases, the gland accumulates MIT and DIT but produces little or no hormone. A recently reported case demonstrating this defect was also associated with congenital deafness, said to be based on a cochlear defect (Hollander *et al.*, 1964).

The deiodinase enzyme has been described as one that salvages iodine from MIT and DIT and makes it available for reutilization in the synthetic pathway. De-

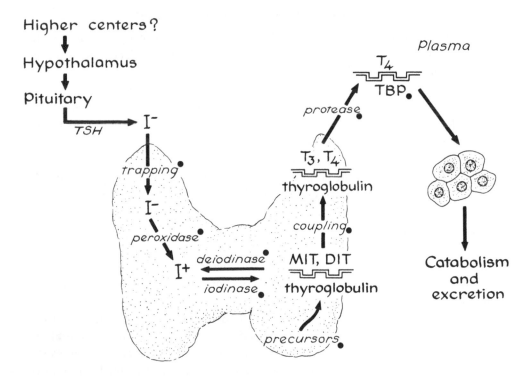

Fig. 1   The Thyroid Pathway.   Possible defects are indicated by •

ficiency of this enzyme allows leakage of iodinated hormone precursors from the thyroid gland and loss in the urine, leading to a depletion of iodine stores and resulting in thyroid hyperplasia.

An abnormal iodinated protein resembling albumin has also been found in association with familial goiter. This substance is recognized by the fact that it is not extractable into butanol. The origin and significance of this compound is not clear at present, though it may be related to a block in the synthesis or degradation of thyroglobulin. Other studies have indicated that thyroglobulin may well be involved in some cases of familial goiter. Van Wyk and his colleagues (1962) reported the pedigree of 23 goiterous individuals in whom a defective thyroglobulin appeared to be present that was more resistant to proteolytic enzymes than normal thyroglobulin. Their data suggest that this trait is inherited as a dominant character.

Another study of a family of goiterous individuals revealed the presence of butanol-soluble iodinated substances in the plasma which resembled partial hydrolytic products of thyroglobulin. The evidence suggests that in this family the defect is in thyroglobulin synthesis, resulting in a molecule that cannot be hydrolyzed normally, and leading to the appearance of incompletely hydrolyzed fragments of thyroglobulin in the plasma (Wynn et al., 1962). A patient has been described in whom this defect as well as the inability to couple iodinated tyrosine molecules was present (Wiener and Lindeboom, 1964); it is suggested that the underlying mechanism in such a case may be the inability to produce normal thyroglobulin.

A deficiency of the thyroglobulin protease enzyme, which is essential to the release of thyroxin from the thyroid, was also reported recently in a person with cretinism (Reinwein and Klein, 1963). Once in the circulating blood, the amount of active thyroxin present is related to the amount of thyroxin-binding protein (TBP). An interesting group of twenty subjects was described recently, of whom ten were clinically hyperthyroid and ten were clinically hypothyroid. All had PBI's within the normal range. A study of the amount of unbound active hormone in these subjects indicated a deficiency of TBP in the patients who appeared hyperthyroid and an excess of TBP in those who were hypometabolic (Lemarchand-Berand and Vannotti, 1963). There is evidence that such variations in TBP have an hereditary basis (Murray and McGirr, 1964; Beierwaltes, 1964). One study indicated that excess TBP may be inherited as a simple Mendelian dominant trait in certain families (Florsheim et al., 1962). Such an individual may have a very high PBI and yet be clinically euthyroid (Ingbar et al., 1964).

The genetic aspects of hyperthyroidism are less clear than those of simple goiter or hypothyroidism. Likewise, the mechanism of hyperthyroidism or thyrotoxicosis is not as well understood. The possibility of a deficiency in thyroxin-binding protein has already been mentioned. An excess of TSH or the presence of an abnormal TSH may be involved in this condition (Purves and Adams, 1963); but very recent reports based on a new TSH assay indicate that TSH levels tend to be somewhat low in established hyperthyroidism. Some genetic studies have indicated a recessive trait in thyrotoxicosis, but others suggest a dominant mode of inheritance (Fraser, 1964). An interesting but not yet understood association exists between the ability to taste phenylthiocarbamide (PTC) and the type of goiter seen in many persons. Persons homozygous for the non-PTC-taster gene have a tendency toward thyroid deficiency, whereas PTC tasters may be more likely to develop thyrotoxicosis (Fraser, 1964). There is some evidence to indicate that the antithyroid antibodies associated with Hashimoto's struma may represent a genetic defect in antibody formation. The mode of inheritance is not clear, but in a few families that have been studied, there has been evidence to support the possibility of dominant transmission (Beierwaltes, 1964).

THYROID-CATECHOL AMINE INTERACTIONS

It has long been recognized that there is a distinct resemblance between the signs of "thyroid storm" and what might be called "adrenergic storm." A great deal of research has gone into the subject of possible interactions between thyroid hormones and catechol amines, which Harrison (1964) discussed in detail in his recent review of this subject. Although such interactions might well be relevant to psychological stress responses, they have scarcely received any consideration in this context.

In an effort to observe the hyperthyroid patient or experimental animal in the absence of adrenergic effects, the techniques of sympathetic block or adrenergic blocking drugs have been utilized. Brewster et al. (1956) studied thyroid-fed dogs before and after sympathetic block produced by epidural procaine injections. (The question of what other effects this block might have has been raised.) He found the elevated heart rate and oxygen consumption produced by thyroid feeding were abolished by the block. He also found that infusions of epinephrine or norepinephrine subsequent to sympathetic block produced a significantly greater increase in heart rate and oxygen consumption in thyroid-fed dogs than in euthyroid animals.

Using a different preparation, Bray (1964) recently reported that surgical removal of the adrenal medulla in rats does not prevent the increase in heart rate and BMR produced by triiodothyronine administration, although the tachycardia could be abolished by reserpine. Although Bray did not find a decrease in BMR with reserpine, other groups (for example, Ramey et al., 1955) have reported prevention of increased oxygen consumption in thyroxin-fed rats with Dibenzyline, an alpha-adrenergic blocking agent.

On the other hand, nethalide, a beta-adrenergic blocking agent, produced no significant changes in heart rate, blood pressure, or oxygen consumption in hyperthyroid patients (Wilson et al., 1964). Guanethidine, a drug that depletes peripheral stores of catechol amines, has been reported to be quite effective in alleviating many of the signs and symptoms of hyperthyroidism. Lee and his colleagues (1962) treated 23 female and four male hyperthyroid patients with guanethidine (average daily dose: 80 mg) and evaluated the patients for changes in general severity of disease, eyelid position, body weight, BMR, blood pressure, pulse pressure, heart rate, and cardiovascular index. They reported significant improvement in all these parameters after an average of fifteen days' treatment, in spite of the fact that there was no change in PBI or I^{131} uptake.

Gaffney and his co-workers (1961) studied many of the same parameters in four normal subjects who were given triiodothyronine. They found that guanethidine lowered the blood pressure, heart rate, and BMR in these subjects, though not all the way to control values. Guanethidine also eliminated the tremor of the outstretched hand that had appeared in these subjects.

Many mechanisms by which thyroid hormones might enhance the effects of catechol amines have been suggested, but these still remain to be clarified.

Swanson (1956) found that epinephrine produced no increase in oxygen consumption in thyroidectomized rats, but an increase in oxygen consumption occurred when epinephrine was combined with thyroxin. He suggested that epinephrine serves to increase the production of substrate by its glycogenolytic action while thyroxin increases the rate of utilization of the glucose. Furthermore, Trendelenburg (1953) found that thyroid feeding of rabbits enhances the hyperglycemic response to epinephrine or norepinephrine; and a more recent study suggested the possibility that this might be due to a thyroxin-induced increase in a phosphorylase (Hornbrook and Brody, 1963).

Various authors have suggested that thyroxin might potentiate the effects of catechol amines by inhibiting in some way the enzymes responsible for their inactivation, namely, monoamine oxidase and catechol-o-methyl transferase. The results of studies in this area appear quite conflicting, probably because different assay techniques, substrates, and tissue sources of the enzymes have been used by the various investigators. It also appears that whatever effect thyroxin may have on these enzymes is not mediated via its acting as a blocking agent, but rather a more subtle effect, for example, inhibiting the synthesis of the enzyme in the liver. Thus, Zile and Lardy (1959) found that thyroid hormones had no effect on MAO activity when added to an *in vitro* preparation, but the livers of rats that had been pretreated with thyroid for several days showed only half the MAO activity as those of untreated controls (using tyramine as substrate). Similarly, D'Iorio and Leduc (1960) found no effect on catechol-o-methyl transferase activity when thyroxin was added to a rat liver homogenate (using protocatechuic acid as substrate). However, there was a significant inhibition of COMT activity in the livers of rats that had been injected with thyroxin for three days prior to testing.

There have been past suggestions of an increased secretion of catechol amines in hyperthyroidism, but this is probably not the case. Ishida (1962) found no significant differences in urinary excretion of adrenaline, noradrenaline, metadrenaline, normetadrenaline, or VMA in 25 hyperthyroid, twenty euthyroid, and two hypothyroid patients. Similar results have been reported by Wiswell and his associates (1963). Studies of catechol amine blood levels and turnover studies in hyperthyroid patients using advanced biochemical techniques would help to clarify this question. Data of this sort under contrasting conditions of emotional distress would be of considerable interest to our present inquiry.

Thyroid-catechol interactions have been approached from another angle. Some experiments have dealt with the effect catechol amines might have on thyroid secretion. Ackerman and Arons (1958) reported an increase in PBI[131] in thyroid vein and femoral artery blood samples in dogs, occurring 15 to 95 minutes after injection of either epinephrine or norepinephrine; but these results are now said to be misleading due to changes in thyroid blood flow and at most involve only a transient fluctuation in secretion (Harrison, 1964). It has also been suggested from time to time that sympathetic nerves to the thyroid might be involved in the regulation of its secretions, but the group of studies by Söderberg (1958) in rabbits and cats indicated that the effects of sympathetic stimulation are primarily those of vasoconstriction within the gland, with questionable slight, transient changes in secretion.

RESEARCH POSSIBILITIES

Almost all research dealing with thyroid stress responses in man has been concerned with thyroid hyperfunction, as we pointed out in the opening section of this paper. This search for *increases* in functional activity of various systems has been a general characteristic of psychosomatic stress research (Hamburg, 1959). It seems to us that there are good reasons for paying attention to *decreases* in functional activity of endocrine glands under stress as well.

In regard to thyroid function, the genetic abnormalities so far delineated (see earlier section) mainly involve some diminution of thyroid hormone secretion. The question thus arises: what happens to individuals who have an unusually limited capacity to synthesize thyroid hormone when they encounter a highly threatening personal situation that is very difficult to resolve? Are the heterozygotes in a position of inadequate functional reserve to meet sustained major stress situations? If so, does the long-term inadequacy of thyroid hormone synthesis have detrimental effects on brain function, such that the solution of the precipitating personal problem becomes even more difficult? The same sorts of questions may well be raised about genetic situations that lead to *excessive* availability of thyroid hormone under major stress conditions; but we are for the moment centering on problems of hypofunction.

One body of information that provides a point of entry into the questions we are raising here is that bearing on the formation of goiter. This is essentially a compensatory hypertrophy of the thyroid in response to its own inadequate hormone synthesis. The moderately low circulating levels of thyroid hormone have feedback effects on the anterior pituitary (perhaps via the hypothalamus, at least in part) such that the secretion of TSH becomes persistently elevated and this gradually produces a thyroid hypertrophy—in the face of the gland's limited synthesizing ability. In time, the thyroid hormone synthesis may become functionally sufficient, but at the price of a palpably enlarged gland.

There is evidence of genetic susceptibility in pathogenesis of goiter. Many individuals who live in hyperendemic goiter areas do not develop goiter; moreover, familial occurrence of simple goiter in nonendemic areas has been observed. There is growing evidence that incomplete defects in thyroid hormone synthesis occur in adult heterozygotes who develop simple goiter without clinical hypothyroidism. For example, a partial deficiency in the enzyme dehalogenase has been discovered in relatives of goitrous cretins, some but not all of whom had goiters. Heterozygotes, having a partial defect, are able to maintain normal thyroid function under usual conditions but develop goiters when faced with extraordinary stresses. Iodine deficiency, pubescence, and pregnancy are the best known situations of this kind. Psychological stresses might well be added to the list. Thus an interaction of genetic predisposing factors and environmental precipitating factors leads to the clinical disorder. Recent genetic evidence indicates that partial defects in synthesis of thyroid hormones are fairly common; that is, the frequency of heterozygotes is much higher than overt thyroid disease. Such partial defects might have no effect on behavior except under severe, prolonged stress.

In a stimulating paper on frontiers in endocrine medicine, Rachmiel Levine made a point that is important in the present context.

In both the adreno-genital syndrome and genetic goiter, the particular enzymatic block must be incomplete, since in the vast majority of cases the hyperplasia of the gland compensates fairly completely for the enzymatic defect. In some instances, however, glucocorticoid insufficiency or myxedema are present in the face of hyperplastic glands (1964, p. 849).

From a psychiatric viewpoint, the question arises whether an inadequate thyroid response to major stress might be significant in the precipitation of some depressive reactions. During a period of bereavement, in which the individual is attempting to work through the problems posed by an important interpersonal loss, it might well be additionally discouraging—indeed, for some people overwhelming—to feel an extraordinary lack of energy, slowness of reaction, and loss of alertness. These latter symptoms, as we have indicated, are frequently observed in clear-cut hypothyroidism. In the present context, we raise the question whether lesser degrees of thyroid limitation might contribute to a sense of inadequacy, failure, or hopelessness at a time when the individual is already faced with a very discouraging situation and coping behavior is being put to a severe test.

In persons whose capacity for synthesis of thyroid hormone is quite limited, the difficulty of coping with stressful circumstances would be compounded if the CNS tended to inhibit thyroid activity. Is there evidence of any such CNS action, or of any psychological conditions under which thyroid activity is diminished? In the course of this paper, we have cited several such observations: (1) a neuroendocrine experiment (Thomas, 1964) indicated that hypothalamic stimulation can elicit inhibition of thyroid function as well as facilitation, the result depending on the placement of the electrodes. This needs replication, but it suggests the existence of a true neuroendocrine *regulatory* system, such as that which has been found for CNS regulation of the adrenal cortex (Hamburg, 1959). (2) In the pioneering study of Brown-Grant, Harris, and Reichlin (1954), several stressors produced a *decrease* in the release of I^{131} labeled hormone from the thyroid of rabbits. Denervation did not block the effect, so evidently something more than vasoconstriction must be involved; presumably the inhibition was mediated via hypothalamic-anterior pituitary-thyroid linkages. (3) In rats, Reiss (1956) found a decrease in I^{131} uptake acutely under two stressful conditions; however, he found the opposite result 24 hours later; in other words, I^{131} uptake was increased in the longer term situation. This calls our attention to the problem of stress-thyroid time relations; we shall return to this problem shortly. (4) In man, Reiss *et al.* (1953) found a tendency to thyroid hypofunction in male anxious patients (in contrast to female anxious patients, who tended to hyperfunction). This too needs replication; sex differences have been neglected in this area of research.

At the very least, one can say that there is a distinct possibility of a neuroendocrine mechanism that inhibits thyroid function and of psychological conditions under which this mechanism tends to operate. This possibility surely deserves investigation.

Thus, both genetically determined enzymatic defects in the thyroid biosynthetic pathway and CNS factors are relevant to thyroid hypofunction under stress. The problem of such hypofunction is perhaps most interesting when viewed in a developmental perspective. We have already pointed out the profound effects of hypothyroidism on brain and behavior in both children and adults. In general, the earlier the hormone deficiency, the more profound and enduring are the behavioral effects.

We have earlier called attention to cretinism as one major indication of the importance of thyroid hormone for behavior. The onset of hypothyroidism in infancy or early childhood is an important cause of mental retardation. Juvenile hypothyroidism, with a later onset, is in many of its characteristics intermediate between cretinism and adult hypothyroidism. There is also the possibility that more subtle variations in the circulating concentration of thyroid hormone— mostly within the "normal range"—may have long-term developmental effects on behavior, through interactions with environmental factors. A statement on thyroid hormone and behavior in Tepperman's excellent account of endocrine and metabolic physiology is relevant here:

One should reflect on the fact that a few micrograms of a rather simple organic chemical substance can cause such profound changes in mood and mentation by its absence or its excessive presence. It seems fair to conclude that variations in the supply of this and other chemical substances within the so-called normal range could help to account at least in part for some of the wide differences in personality structure seen in the "healthy" ambulatory population. . . . The capacity to solve "problems in living" may depend, to some extent, on the nature of one's inborn endocrine equipment (1962, p. 98).

One particularly challenging aspect of the line of inquiry we are now pursuing is that it might well involve a conjunction between modern biochemical-genetic-endocrine techniques and those of modern personality development dealing with interpersonal-socioenvironmental variables. By way of illustration, let us consider an hypothetical situation. Suppose a child is heterozygous for a gene responsible for one of the defects in thyroid hormone synthesis, and therefore quite limited in his capacity to respond with increased thyroid hormone production even in the face of major CNS-thyrotropin stimulation. Now, further suppose that such a child encounters a personal crisis of intense and prolonged character, such as might be precipitated by the loss of a significant person in the child's life. What happens under these conditions? It is possible that the full-blown picture of juvenile hypothyroidism might be precipitated. But we wonder whether, more commonly, the result might be borderline-low levels of circulating thyroid hormone over an extended period, leading to moderate (often subclinical) effects on behavior. These effects might well include a tendency in any of the following directions: slowing of intellectual activity; decrease in emotional responsiveness; drowsiness; sensitivity to cold; greater susceptibility to fatigue and to infectious diseases. Any changes of this sort in the child might well evoke significant reactions from parents, siblings, friends, or teachers. The ultimate course of the child's personality development could be influenced by the reaction of a highly significant adult to such a change in the child's behavior. If, for example, such a response were strongly depreciatory in character—for example, if the child's behavior changes were taken as indications of stupidity or general inadequacy—the long-term results might be quite important. Moreover, the child's long-term style of coping with stressful events could be shaped by experiences of this sort.

The application of a developmental viewpoint to thyroid *hyperfunction* is also of interest. Here again, a subclinical change in circulating concentration of thyroid hormone—in this case, a sustained, moderate *elevation*—might well produce behavioral changes that would be taken seriously by parents and other significant persons in the child's life. The well-known effects of thyroid hyperfunction on behavior include: impatience, irritability; restlessness, hyperactivity; anxiety-proneness; extended wakefulness, difficulty in sleeping; sensitivity to heat. It is not difficult to imagine that different parents might place quite different interpretations on such behavior. Some mothers would be highly threatened by such a child, perceiving him as difficult, demanding, and hard to manage. In general, behavior of this kind would (in many societies) be taken as more acceptable in a boy than in a girl—part of the vigorous, assertive disposition expected of boys. Indeed, some parents (especially fathers who are not faced with the hour-to-hour problems of behavior regulation) might take pride in such behavior in a son as indicating a particularly masculine disposition.

Here we are raising the possibility of complex interactions between CNS and thyroid specifically, and genetic and environmental factors generally. Genetically determined protein abnormalities in the transport or disposal of thyroid hormones are capable of producing a tendency toward elevation in circulating thyroid hormones—and any such tendency would be accentuated under conditions of enhanced secretion, such as might occur under prolonged psychological stress. Then, the behavior changes induced by prolonged (though perhaps moderate) elevations in circulating thyroid hormone would evoke reactions from persons significant to the child—and might in time affect the child's view of himself. While these considerations are necessarily speculative, we hope they will provide an impetus to research. We believe that recent advances make possible the analysis of such interactions.

This brings us to a brief consideration of promising research directions in regard to the bearing of psychological factors upon thyroid *hyperfunction*. The clinical observations, and some of the systematic studies of human subjects, are sufficiently impressive that the problem deserves pursuit with the best available techniques. Yet, as we have pointed out, the literature is ambiguous in important respects. An additional element of complexity is added by very recent findings, still largely unpublished, that have come to our attention just before completion of this paper (Kriss, personal communication). With newer techniques of TSH assay, it appears that TSH is not elevated in adult hyperthyroidism, at least in most of the established cases that have so far been studied. It may turn out that only one subgroup of hyperthyroid patients shows an elevated TSH. Indeed, it has been suggested (Tepperman, 1962) that hyperthyroidism may well turn out to be heterogeneous. This has been a rather general finding of genetically oriented clinical research in recent years.

In regard to behavior, the heterogeneity of mental retardation provides the best known situation. Clearly, a variety of genetic-biochemical-environmental pathways may lead to mental retardation. Similarly, broad diagnostic categories applying to common diseases such as depression, schizophrenia, hypertension, and hyperthyroidism probably cover a variety of subgroups. Differentiation of such subgroups on clinical, behavioral, biochemical, or genetic grounds is likely to provide a profitable route of research. The relevance of this approach to psychiatric and psychosomatic problems is developed more fully elsewhere (Hamburg, 1966). Lloyd (1963) provides an example of such a subgroup in thyroid stress responses. In a stimulating review of human neuroendocrinology, he points out that there are hyperthyroid patients who "have a tendency to favor TSH, rather than ACTH, secretion under emotional stress." The availability of immunochemical methods for measuring pituitary tropic hormones within the next few years will provide an opportunity for direct assessment of this problem.

Gibson (1962), in a review of the stress-thyroid literature, is not impressed with psychological stress effects on thyroid function in general. However, he qualifies this conclusion by pointing out that such effects may become significant for the individual if (1) there is a family history of hyperthyroidism; and (2) the stress is operative over an extended period of time. This view is quite consistent with our own experience and with the views of several clinical endocrinologists who have taken an interest in this problem. The following quotation from Rulon Rawson provides a useful illustration of the latter viewpoint.

Graves' disease is usually characterized by exophthalmos, goiter, and thyrotoxicosis. The cause is unknown. However, there is certainly a genetic predisposition. Moreover, the relatives of patients suffering from Graves' disease may show alteration in thyroid function even though they are euthyroid at the time of examination. For example, in one series 50 percent of the relatives of Graves' disease patients had an increase in the uptake of I^{131}. In these individuals, the net euthyroid status was due to an increase in the degradation as well as in the production of thyroxine.

One cannot help but wonder, therefore, whether Graves' disease may not be basically a familial condition, in which the full-blown clinical picture is merely triggered into existence by stress—emotional or physical (1965, p. 47).

This view has research implications. We are impressed with the desirability of identifying predisposed individuals and following them longitudinally through periods of major stress. For instance, one might pick subjects from a population at risk for stressful experience within the foreseeable future. These subjects would have no overt thyroid disease, but would have one or more of the following characteristics (or other similar ones): (1) family history of hyperthyroidism (or, more broadly, of thyroid disease); (2) BEI in upper portion of the "normal range"; (3) moderately elevated I^{131} uptake. Thus, one would have subjects who, though clinically euthyroid, were presumably predisposed toward thyroid hyperfunction. If any persons are likely to demonstrate thyroid hyperfunction under stress, these are the ones. In view of Reiss's data (cited earlier) indicating that women rather than men show a tendency toward thyroid hyperfunction when anxious, the study of female subjects might provide further loading in this direction. Under these conditions, it would be possible to use each individual as her own control in comparing high-distress and low-distress states and their concomitant thyroid responses. Also, the duration of distress required to elicit significant thyroid hyperfunction could be assessed, both intra- and inter-individually.

In reviewing the stress-thyroid literature, it is apparent that the relevant time relations have not received the attention they deserve. Future studies will do well to specify—and to vary systematically—the *duration* of emotional distress. Moreover, the time course of thyroid function is of much interest—before, during, and after the stressful period, to the extent that such measures are possible.

When a single dose of thyrotropin is given experimentally, the *release of stored hormones* from the colloid is apparent within thirty minutes. When I^{131} has previously been given, an acceleration of I^{131} release from the gland is measurable within thirty minutes after intravenous injection of thyrotropin (Purves, 1964). This rapid release of stored hormone under thyrotropin stimulation may account for the observations in several studies (Hetzel *et al.*, 1952, 1956; Alexander *et al.*, 1961) indicating enhancement of thyroid function immediately following an acute stressful experience. A study reported quite recently (Falconer and Hetzel, 1964) is interesting in this context. An exteriorized thyroid gland preparation in sheep permitted collection of thyroid vein blood in waking, undisturbed animals. Rises in PBI^{131} and PBI were detected fifteen to thirty minutes after insertion of a cannula into the jugular vein. When this response had subsided, similar rises occurred after a series of fireworks explosions.

Finally, and most consistently, rises in PBI^{131} and PBI were elicited by exposure of the sheep to a barking dog. These rises lasted up to two hours. Restraint produced a similar effect, which disappeared after training. Similar but larger PBI and PBI[131] increases were elicited with a large dose of thyrotropin. Incidentally, the restraint finding calls attention to yet another source of variation: species differences. There appears to be marked species variation in thyroid responses to restraint; for example, inhibition in rabbits, stimulation in sheep. This species factor is worth considering in future experiments. In any event, the utilization of similar preparations with direct access to thyroid vein blood in waking animals would be of much interest in *chronic* experiments, especially in primates.

In regard to longer-term stimulation of the gland, the situation becomes more complex and probably more relevant to the clinical problems of major stress. Purves's cogent summary of the experimental evidence will provide the background information for the point we wish to make:

> Over an extended period the rate of secretion by the thyroid gland is limited to the rate at which hormonally active iodine compounds are synthesized within it. The incorporation of iodine into thyroglobulin is a process requiring chemical energy, and one moreover which can proceed only *pari passu* with the synthesis of the protein itself. Acceleration of iodine assimilation in response to stimulation by thyrotrophin is subject to a considerable lag and continues for a considerable time after thyrotrophin is withdrawn. From the fact that increasing stimulation leads to increasing depletion of the iodine content of the thyroid gland it is clear that iodine assimilation is accelerated to a lesser degree than hormone release. At high degrees of stimulation continued for an extended period, the rate of hormone synthesis rather than the ability to mobilize stored hormone becomes the factor determining the rate of secretion (1964, pp. 5–6).

These facts suggest the following sequence in thyroid response to intense, sustained stress: an initial, transient elevation of circulating hormone concentration due to release of stored hormone; a period of somewhat lower levels, reflecting the lag in new synthesis; and finally (as the stress continues), high levels reflecting new synthesis of hormone. The latter situation provides an interesting analogy with major stress situations (of clinical significance) that extend over many days, weeks, or even months. Intense, *prolonged* distress may provide—at least in some individuals—a chronic CNS-thyrotropin-thyroid stimulation. This view is consistent with the emphasis on long-term stress in Gibson's (1962) review and with several of the more rigorous human studies cited earlier in this paper.

Throughout this paper, we have attempted to call attention to gaps in information, newer research opportunities, and promising lines of inquiry. For example, the potent interactions of thyroid hormones and catechol amines are quite intriguing for stress research, since there is so much evidence of increase in catechol amine secretion under stress. In the presence of high catechol amine production —as in intense anxiety—even modest increases in circulating thyroid hormone might have striking effects, for example, on CNS and behavior. We do not wish to repeat the various research implications which we have mentioned earlier, but to only close by reiterating our belief that the conjunction of genetic-biochemical techniques with newer approaches to a scientific study of behavior will prove rewarding in the problem area of this paper. (In addition to the papers already cited, several valuable reviews were recently published: Harris, 1964; Reichlin, 1964; Michael and Gibbons, 1963; Williams and Baake, 1962; Werner, 1961.)

SUMMARY

In an analysis of the clinical and experimental literature on thyroid-brain relations, an integrated behavior-endocrine-genetic approach to stress problems is presented. This includes a review and analysis of the research literature on the following topics: (1) changes in thyroid function associated with emotional distress; (2) changes in thyroid function associated with personality characteristics and psychiatric disorders; (3) biosynthesis, secretion, and catabolism of thyroid hormones; (4) assessment of thyroid function; (5) CNS regulation of thyroid function; (6) effects of thyroid hormones and of abnormalities in thyroid function on behavior; (7) effects of thyroid hormones and abnormalities in thyroid function on development of the brain and development of behavior; (8) genetically determined abnormalities in thyroid function; (9) interactions of thyroid hormones and catechol amines. Throughout this paper, and particularly in the final section, emphasis is placed upon specific lines of inquiry that appear promising in light of recent advances in biological and behavioral sciences.

BIBLIOGRAPHY

ACKERMAN, N. B., and W. L. ARONS
   1958. The effect of epinephrine and norepinephrine on the acute thyroid release of thyroid hormones. *Endocrin.*, 62:723.

ALEXANDER, FRANZ, *et al.*
   1961. Experimental studies of emotional stress. *Psycho. Med.*, 22:104.

ANDERSSON, B., *et al.*
   1962. Thyroidal response to local cooling of the pre-optic "heat loss center." *Life Sci.*, 1:1.

ARTUNKAL, S., and B. TOGROL
   1964. Psychological studies in hyperthyroidism. In *Brain-thyroid relationships*, CIBA Foundation Study Group No. 18. Boston: Little, Brown.

ASHER, R.
   1949. Myxedematous madness. *Brit. Med. J.*, 2:555.

BARANOV, G. G., *et al.*
   1961. Nervous factors in the pathogenesis of thyrotoxicosis. In R. Pitt-Rivers (Ed.), *Advances in thyroid research*. New York: Pergamon Press.

BASCHIERI, L., *et al.*
   1963. Modifications of thyroid activity by melantonin. *Experientia*, 19:15.

BEIERWALTES, W. H.
   1964. Genetics of thyroid disease. In *The thyroid*, International Academy of Pathology Monograph No. 5. Baltimore: Williams and Wilkins.

BEIERWALTES, W. H., *et al.*
   1959. Institutionalized cretins in the state of Michigan. *J. Mich. Med. Soc.*, 58:1077.

BENNETT, A. W., and C. G. CAMBOR
1961. Clinical study of hyperthyroidism: a comparison of male and female characteristics. *Arch. Gen. Psych.*, 4:160.

BOARD, F., H. PERSKY, and D. A. HAMBURG
1956. Psychological stress and endocrine functions: blood levels of adrenocortical and thyroid hormones in acutely disturbed patients. *Psycho. Med.*, 18:324.

BOARD, F., R. WADESON, and H. PERSKY
1957. Depressive affect and endocrine functions. *Arch. Neurol. Psych.*, 78:612.

BONETTI, U.
1958. The problem of mental syndromes in hyperthyroidism and after thyroidectomy. *Cervello* (Naples), 34:305.

BOWMAN, K. M., et al.
1950. Thyroid function in mental disease. *J. Nerv. Ment. Dis.*, 112:404.

BRAY, G. A.
1964. Studies on the interactions of thyroid hormone and catechol amines. *J. Clin. Invest.*, 43:285.

BREWSTER, W. R., et al.
1956. The hemodynamic and metabolic interrelationships in the activity of epinephrine, norepinephrine, and the thyroid hormones. *Circulation*, 13:1.

BRODEMAN, D. D., and R. M. WHITMAN
1952. Post-thyroidectomy psychoses. *J. Nerv. Ment. Dis.*, 116:340.

BRODY, E. B., and E. B. MAN
1950. Thyroid function measured by serum precipitable iodine determinations in schizophrenic patients. *Amer. J. Psych.*, 107:357.

BROWN-GRANT, K., G. W. HARRIS, and S. REICHLIN
1954. The effect of emotional and physical stress on thyroid activity in the rabbit. *J. Physiol.*, 126:29.

1957. Effect of pituitary stalk section on thyroid function in the rabbit. *J. Physiol.*, 136:364.

BURKLE, J. S., and R. LUND
1963. Measurements of thyroid function with various radioisotopes of iodine. In F. W. Sunderman and F. W. Sunderman, Jr. (Eds.), *Evaluation of Thyroid and Parathyroid Functions*. Philadelphia: J. B. Lippincott.

CARPELAN, P. H.
1953. Psychological abnormalities following thyroidectomy. *Acta Psych. et Neurol.* (Copenhagen) Suppl., 80:217.

CHAPTL, J., et al.
1956. Cerebral involvement in infantile myxedema: an electroencephalographic study. *Presse Medical* (Paris), 64:2257.

D'ANGELO, S. A.
1963. Central nervous regulation of the secretion and release of thyroid stimulating hormone. In A. V. Nalbandov (Ed.), *Advances in Neuroendocrinology*. Urbana: University of Illinois Press.

D'ANGELO, S. A., and J. SNYDER
1963. Electrical stimulation of the hypothalamus and TSH secretion in the rat. *Endocrin.*, 73:75.

D'ANGELO, S. A., et al.
1964. Electrical stimulation of the hypothalamus: simultaneous effects on the pituitary-adrenal and thyroid systems of the rat. *Endocrin.*, 75:417.

DANZIGER, L.
1958. Thyroid therapy of schizophrenia. *Dis. Nerv. Syst.*, 19:373.

DANZIGER, L., and J. KINDWALL
1948. Thyroid therapy in some mental disorders. *Dis. Nerv. Syst.*, 9:231.

DEGROOT, L. J.
1960. Thyroid function tests. In L. B. Page and P. J. Culver (Eds.), *A syllabus of laboratory examinations in clinical diagnosis.* Cambridge, Mass.: Harvard University Press.

DE JONG, W., and J. MOLL
1965. Differential effects of hypothalamic lesions on pituitary-thyroid activity in the rat. *Acta Endocrin.*, 48:522.

DEVISSCHER, M., et al.
1964. Is there any fundamental difference between endemic and sporadic non-toxic goitre? *Acta Endocrin.*, 45:365.

DIGRUTTOLA, G.
1958. Electroencephalographic observations in the course of infantile myxedema. *Pediatria* (Naples), 66:981.

D'IORIO, A., and J. LEDUC
1960. The influence of thyroxine on the O-methylation of catechols. *Arch. Biochem. and Biophys.*, 87:224.

DONGIER, M., et al.
1956. Psychophysiological studies in thyroid function. *Psycho. Med.*, 18:310.

EAYRS, J. T.
1960. Influence of the thyroid on the central nervous system. *Brit. Med. Bull.*, 16:122.

EAYRS, J. T., and W. A. LISHMAN
1955. The maturation of behavior in hypothyroidism and starvation. *Brit. J. Animal Behav.*, 3:17.

EVANS, E. C.
1960. Neurological complications of myxedema. *Ann. Int. Med.*, 52:434.

FALCONER, I. R., and B. S. HETZEL
1964. Effect of emotional stress and TSH on thyroid vein hormone level in sheep with exteriorized thyroids. *Endocrin.*, 75:42.

FELDMEYER-REISS, E. E.
1958. Application of triiodothyronine in the treatment of mental disorders. *J. Nerv. Ment. Dis.*, 127:540.

FERRIER, P. E., and T. LEMARCHAND-BERAND
1965. Variations of thyroid hormone transport in the blood of normal children and of children with mongolism. *Acta Endocrin.*, 48:547.

FLORSHEIM, W. H., et al.
1962. Familial elevation of serum thyroxine-binding capacity. *J. Clin. Endocrin. and Metab.*, 22:735.

FORD, D. H., and J. GROSS
1958a. The metabolism of I[131] labeled thyroid hormones in the hypophysis and brain of the rabbit. *Endocrin.*, 62:416.

1958b. The localization of I[131] labeled triiodothyronine and thyroxine in the pituitary and brain of the male guinea pig. *Endocrin.*, 63:549.

FORD, D. H., et al.
1959. The localization of I[131] labeled triiodothyronine in the pituitary and brain of normal and thyroidectomized rats. *Endocrin.*, 64:977.

1964. Further studies on differences in the uptake of I^{131}-triiodothyronine in the brains of male and female rats. *Acta Endocrin.*, 45:219.

FOTOPULOS, D.
1958. Alternating hemiplegia in myxedema. *Psych., Neurol., und Mediz. Psych.* (Leipzig), 10:66.

FRASER, G. R.
1964. Genetical aspects of thyroid disease. In R. Pitt-Rivers and W. R. Trotter (Eds.), *The thyroid gland*, Vol. 2. Washington, D.C.: Butterworths.

GAFFNEY, T. E., *et al.*
1961. Effects of guanethidine on tri-iodothyronine-induced hyperthyroidism in man. *N. Eng. J. Med.*, 265:16.

GARCIA, J., G. W. HARRIS, and W. J. SCHINDLER
1964. Vasopressin and thyroid function in the rabbit. *J. Physiol.*, 170:487.

GELBER, S., *et al.*
1964. Effects of L-thyroxine on amino acid incorporation into protein in mature and immature rat brain. *J. Neurochem.*, 11:221.

GIBSON, J. G.
1962. Emotions and the thyroid gland: a critical appraisal. *J. Psycho. Res.*, 6:93.

GJESSING, L. R.
1964. Studies of periodic catatonia. 1. Blood levels of protein-bound iodine and urinary excretion of vanillyl-mandelic acid in relation to clinical course. *J. Psychiat. Res.*, 2:123.

GJESSING, R.
1938. Disturbances of somatic functions in catatonia with a periodic course and their compensation. *J. Ment. Sci.*, 84:608.

GOOLDEN, A. W. G.
1960. Use of radioactive iodine in the diagnosis of thyroid disorders. *Brit. Med. Bull.*, 16:105.

GREER, M. A.
1963. Supra-hypothalamic control of thyrotropin secretion. In S. C. Werner (Ed.), *Thyrotropin*. Springfield, Ill.: Charles C Thomas.

GREGORY, IAN
1956. Mental disorder associated with thyroid dysfunction. *Canad. Med. Assoc. J.*, 75:489.

GUILLEMIN, R., *et al.*
1962. Presence of a substance in the hypothalamus which stimulates the secretion of TSH. *C. R. Acad. Sci.* (Paris), 255:1018.

1963. *In vitro* secretion of thyrotropin (TSH): stimulation by a hypothalamic peptide (TRF). *Endocrin.*, 73:564.

HALASZ, B., *et al.*
1963. Changes in the pituitary-target gland system following electrolytic lesion of median eminence and hypophyseal stalk in male rats. *Acta Morph. Acad. Sci. Hung.*, 12:23.

HAM, G. D., F. ALEXANDER, and H. CARMICHAEL
1951. A psychosomatic theory of thyrotoxicosis. *Psycho. Med.*, 13:18.

HAMBURG, D. A.
1959. Some issues in research on human behavior and adrenocortical function. *Psychosom. Med.*, 21:387.

1966. Genetics of adrenocortical hormone metabolism in relation to psychological stress. In J. Hirsch (Ed.), *Behavior-Genetic Analysis*. New York: McGraw-Hill.

HAMBURGH, M., and R. P. BUNGE
1964. Evidence for a direct effect of thyroid hormone on maturation of nervous tissue grown *in vitro*. *Life Sci.*, 3:1423.

HAMILTON, J. A.
1962. *Postpartum psychiatric illness*. St. Louis: Mosby.

HARRIS, G. W.
1962. Neuroendocrine relations. In S. Korey (Ed.), *Ultrastructure and metabolism of the nervous system*. Baltimore: Williams and Wilkins.

1964. A summary of some recent research on brain-thyroid relationships. In *Brain-thyroid relationships*, CIBA Foundation Study Group No. 18. Boston: Little, Brown.

HARRIS, G. W., and J. W. WOODS
1958. The effect of electrical stimulation of the hypothalamus or pituitary gland on thyroid activity. *J. Physiol.*, 143:246.

HARRISON, T. S.
1964. Adrenal medullary and thyroid relationships. *Phys. Rev.*, 44:161.

HERMANN, H. T., and G. G. QUARTON
1965. Psychological changes and psychogenesis in thyroid hormone disorders. *J. Clin. Endocrin.*, 25:327.

HETZEL, B. S., et al.
1952. Rapid changes in plasma PBI in euthyroid and hyperthyroid subjects. *Trans. Am. Goiter Ass.*, 242.

1956. Changes in urinary nitrogen and electrolyte excretion during stressful life experiences and their relation to thyroid function. *J. Psycho. Res.*, 1:177.

HINMAN, F. J., et al.
1957. Observations upon some relationships between emotional traumata and hyperthyroidism. Paper presented at Western Divisional Meeting of American Psychological Association, Los Angeles, 1957.

HOLLANDER, C. S., et al.
1964. Congenital deafness and goiter. *Am. J. Med.*, 37:630.

HORNBROOK, K R., and T. BRODY
1963. The effect of catechol amines on muscle glycogen and phosphorylase activity. *J. Pharm. Expt. Therap.*, 140:295.

HUBER, G.
1956. Psychoses in late acquired hypothyroidism and hypoparathyroidism. *Nervenarzt*, 27:440.

IMARISIO, J. J., and J. GRECO
1964. Thyroid hormone binding by serum proteins. *Metabolism*, 13:897.

INGBAR, S. H., et al.
1964. Observations on the nature of the underlying disorder and the occurrence of associated plasma transport abnormalities in a patient with an idiopathic increase in the plasma thyroxine-binding globulin. *J. Clin. Invest.*, 43:2266.

ISHIDA, N.
1962. Metabolism of catecholamines in hyper- and hypo-thyroid patients. *Tohaku J. Exper. Med.*, 78:228.

JELLINEK, E. H., and R. E. KELLY
1960. Cerebellar syndrome in myxedema. *Lancet*, 2:225.

JONES, K.
1959. Admission to mental hospital after thyroidectomy—observations on a series of cases. *J. Ment. Sci.*, 105:803.

KELSEY, F., and A. GULLOCK
1957. Thyroid activity in hospitalized psychiatric patients: relation of dietary iodine to I^{131} uptake. *Arch. Neurol. Psych.*, 77:543.

KITCHIN, F. D., and W. H. EVANS
1960. Genetic factors in thyroid diseases. *Brit. Med. Bull.*, 16:148.

KLEINSCHMIDT, H. J., et. al.
1956. Psychophysiology and psychiatric management of thyrotoxicosis. *J. Mt. Sinai Hosp.*, 23:131.

KRACHT, J.
1954. Fright toxicosis in the wild rabbit: a model thyrotropic alarm reaction. *Acta Endocrin.*, 15:355.

KRISS, J. P.
1964. Unpublished lecture. Stanford University School of Medicine.

KRISS, J. P., et al.
1964. Isolation and identification of the long-acting thyroid stimulator and its relation to hyperthyroidism and circumscribed pretibial myxedema. *J. Clin. Endocrin.*, 24:1005.

LANSING, R. W., and J. B. TRUNNELL
1963. Electro-encephalographic changes accompanying thyroid deficiency in man. *J. Clin. Endocrin.*, 23:470.

LEE, W. Y., et al.
1962. Studies of thyroid and sympathetic nervous system interrelationships. II. Effects of guanethidine on manifestations of hyperthyroidism. *J. Clin. Endocrin.*, 22:879.

LEMARCHAND-BERAUD, T., and A. VANNOTTI
1963. Alteration of the plasma transport of thyroid hormone in human pathology. *Schweiz. Med. Wschr.*, 93:7.

LESZYNSKY, H. E.
1964. Genetic studies in familial goitrous cretinism. *Acta Endocrin.*, 46:103.

LEVINE, R.
1964. Selected problems in endocrine medicine. In G. Pincus (Ed.), *The hormones*, Vol. 5. New York: Academic Press.

LIDZ, THEODORE
1949. Emotional factors in etiology of hyperthyroidism. *Psycho. Med.*, 11:2.

LLOYD, C. W.
1963. Central nervous system regulation of endocrine function in the human. In A. V. Nalbandov (Ed.), *Advances in Neuroendocrinology*. Urbana: University of Illinois Press.

LUBART, J. M.
1964. Implicit personality disorder in patients with toxic and non-toxic goiter. *J. Nerv. Ment. Dis.*, 138:255.

LUPULESCU, A., et al.
1962. Neural control of the thyroid gland: studies on the role of extra-pyramidal and rhinencephalon areas in the development of the goiter. *Endocrin.*, 70:517.

MAN, E. B.
1963. Butanol-extractable iodine in serum. In F. W. Sunderman and F. W. Sunder-

man, Jr. (Eds.), *Evaluation of Thyroid and Parathyroid Functions*. Philadelphia: J. B. Lippincott.

Matsuda, K., M. A. Greer, and C. Duyck
1963. Neural control of thyrotropin secretion: Effect of forebrain removal on thyroid function. *Endocrin.*, 73:462.

Means, J. H.
1948. *The thyroid and its diseases*. Philadelphia: J. B. Lippincott.

Mess, B.
1958. Veranderungen des Gehaltes der Hypophyse an thyreotrophen Hormon nach Thyroidektomie unter gleichzeitiger Laison der Nuclei Habenuale. *Endokrin.*, 35:296.

1959. Die Rolle der Nuclei habenulae bei der auf erhöten Thyroxin-Blutspiegel eintretenden zentral-nervosen Hemmung der thyreotrophen Aktivität des Hypophysenvorderlappens. *Endokrin.*, 37:104.

Michael, R. P., and J. L. Gibbons
1963. Interrelationships between the endocrine system and neuropsychiatry. *Int. Rev. Neurobiol.*, 5:243.

Mittlemann, B.
1933. Psychogenic factors and psychotherapy in hyperthyreosis and rapid heart imbalance. *J. Nerv. Ment. Dis.*, 77:465.

Money, J.
1956. Psychologic studies in hypothyroidism. *Arch. Neurol. Psych.*, 76:296.

Murray, J. P. C., and E. M. McGirr
1964. Iodine metabolism in thyroid dysfunction. In R. Pitt-Rivers and W. R. Trotter (Eds.), *The Thyroid Gland*, Vol. 2. Washington: Butterworths.

Nemeth, S., and J. Ruttkay-Nedecky
1958. The etiology of thyrotoxicosis. *Ceskoslovenska Neurologic* (Prague), 21:354.

Nieman, E. A.
1959. EEG in myxedema coma: clinical and EEG study of 3 cases. *Brit. Med. J.*, 2:1204.

Parhon-Stefanescon, K., and A. Vrejon
1957. Thyroidectomy as a therapeutic procedure in affective states. *Z. Nevropatologic I Psihitric. Imeni S. S. Korsakova* (Moscow), 57:1005.

Peters, J. P., and E. B. Man
1950. The significance of serum cholesterol in thyroid disease. *J. Clin. Invest.*, 29:1.

Pitt-Rivers, R.
1960. Biosynthesis of the thyroid hormones. *Brit. Med. Bull.*, 16:118.

Purves, H. D.
1964. Control of thyroid function. In R. Pitt-Rivers and W. R. Trotter (Eds.), *The Thyroid Gland*, Vol. 2. Washington: Butterworths.

Purves, H. D., and D. D. Adams
1963. The long-acting thyroid stimulator in the serum of patients with Graves disease. In S. C. Werner (Ed.), *Thyrotropin*. Springfield, Ill.: Charles C Thomas.

Radcliff, F. J., et al.
1964. Diagnostic value of the estimation of proteinbound iodine in thyroid disease: survey of an Australian population group. *J. Clin. Endocrin.*, 24:883.

Ramey, E. R., et al.
1955. Effect of sympathetic blocking agents on the increased oxygen consumption following administration of thyroxine. *Fed. Proc.*, 14:118.

RAWSON, R. W.

1965. The thyroid gland. *Clinical Symposia*, 17:35–63.

REICHLIN, S.

1963a. Neuroendocrinology. *N. Eng. J. Med.*, 269:1296.

1963b. Regulation of pituitary thyrotropin release. In S. C. Werner (Ed.), *Thyrotropin*. Springfield, Ill.: Charles C Thomas.

1964. Function of the hypothalamus in regulation of pituitary-thyroid activity. In *Brain-thyroid relationships*, CIBA Foundation Study Group No. 18. Boston: Little, Brown.

REICHLIN, S., *et al.*

1963. A critical evaluation of the "TRF" of Shibusawa. *Endocrin.*, 72:334.

REINWEIN, D., and E. KLEIN

1963. A special form of faulty iodine utilization in sporadic cretinism. *Schweiz. Med. Wschr.*, 93:1213.

REISS, M.

1956. Endocrine concomitants of certain physical psychiatric treatments. *Internat. Record Med.*, 169:431.

REISS, M., *et al.*

1953. The significance of the thyroid in psychiatric illness and treatment. *Brit. Med. J.*, 1:906.

RIVLIN, R. S., K. L. MELMON, and A. SJOERDSMA

1965. An oral tyrosine tolerance test in thyrotoxicosis and myxedema. *New Eng. J. Med.*, 272:1143.

ROBBINS, J.

1963. Method for measurement of thyroxine-binding proteins in serum. In F. W. Sunderman, and F. W. Sunderman, Jr. (Eds.), *Evaluation of thyroid and parathyroid functions*. Philadelphia: J. B. Lippincott.

ROBBINS, L. R., and D. B. VINSON

1960. Objective psychologic assessment of the thyrotoxic patient and the response to treatment: preliminary report. *J. Clin. Endocrin. Metab.*, 20:120.

RULLO, F. R., and F. N. ALLAN

1958. Psychoses resulting from myxedema. *J.A.M.A.*, 168:890.

SAXENA, K. M., J. D. CRAWFORD, and N. B. TALBOT

1964. Childhood thyrotoxicoses: a long-term perspective. *Brit. Med. J.*, 2:1153.

SCHEINBERG, P., *et al.*

1950. Correlative observations on cerebral metabolism and cardiac output in myxedema. *J. Clin. Invest.*, 29:1139.

SCHREIBER, V., *et al.*

1962. Isolation of a hypothalamic peptide with TRF (thyreotropin releasing factor) activity in vitro. *Experientia*, 18:338.

SENSENBACH, W., *et al.*

1954. The cerebral circulation in hyperthyroidism and myxedema. *J. Clin. Invest.*, 33:1434.

SHERMAN, L., *et al.*

1963. The Achilles reflex: a diagnostic test of thyroid dysfunction. *Lancet*, 1:243.

SHIBUSAWA, K., *et al.*

1956. The hypothalamic control of the thyrotroph-thyroidal function. *Endocrin. Japon.*, 3:116.

SHIZUME, K., *et al.*

1962. Effect of electrical stimulation of the limbic system on pituitary-thyroidal function. *Endocrin.*, 71:456.

SIMPSON, G. M., *et al.*

1964. Thyroid indices in chronic schizophrenia. *J. Nerv. Ment. Dis.*, 138:581.

SÖDERBERG, U.

1958. Short-term reactions in the thyroid gland. *Acta Physiol. Scand.*, 42: Supp. 147.

SMETTERS, G. W.

1963. The present status of the basal metabolic rate. In F. W. Sunderman and F. W. Sunderman, Jr. (Eds.), *Evaluation of thyroid and parathyroid function*. Philadelphia: J. B. Lippincott.

STANBURY, J. B.

1963. The metabolic errors in certain types of familial goiter. *Rec. Prog. Horm. Res.*, 19:547.

STERLING, K, and R. B. CHODOS

1956. Radiothyroxine turnover in myxedema, thyrotoxicosis, and hypermetabolism without endocrine disease. *J. Clin. Invest.*, 35:806.

STERN, M. H.

1956. Effect of anti-thyroid therapy on objective test performance. *Canad. J. Psych.*, 10:226.

STRISOWER, E. H., *et al.*

1958. Physiologic effects of 1-triiodothyronine. *J. Clin. Endocrin. Metab.*, 18:721.

SUNDERMAN, F. W., JR.

1963. The measurement of serum protein-bound iodine. In F. W. Sunderman and F. W. Sunderman, Jr. (Eds.), *Evaluation of thyroid and parathyroid function*. Philadelphia: J. B. Lippincott.

SWANSON, H. E.

1956. Interrelations between thyroxin and adrenalin in the regulation of oxygen consumption in the albino rat. *Endocrin.*, 59:217.

TEPPERMAN, J.

1962. *Metabolic and endocrine physiology*. Chicago: Year Book Medical Publishers. Chapter 6.

THOMAS, S.

1964. Hypothalamic regulation of thyroid activity. *Indian J. Physiol. Pharm.*, 8:125.

THOULD, A. K., and E. F. SCOWEN

1964. The syndrome of congenital deafness and simple goitre. *J. Endocrin.*, 30:69.

TOLAN, E. J., and L. DILLON

1960. Treatment of chronic schizophrenics with 1-iodothyronine. *Amer. J. Psych.*, 116:1110.

TRENDELENBURG, U.

1953. Thyroid and hyperglycaemia produced by adrenaline and noradrenaline. *Brit. J. Pharm.*, 8:454.

TURMEL, J., *et al.*

1956. Psychiatric aspect of hypothyroidism. *Laval Med.* (Paris), 21:295.

UOTILA, U. U.

1939. On the role of the pituitary stalk in the regulation of the anterior pituitary, with special reference to the thyrotropic hormone. *Endocrin.*, 25:605.

1940. Hypothalamic control of anterior pituitary function. *Res. Pub. Ass. Nerv. Ment. Dis.*, 20:580.

VAGUE, J., et al.

1958. The practical value of the EEG in the study of dysthyroid syndromes. *Gazette des Hopitaux Civils et Militaires* (Paris), 12:485.

VAN WYK, J. J., et al.

1962. Genetic studies in a family with "simple" goiter. *J. Clin. Endoc. Metab.*, 22:399.

VOLPE, R., et al.

1960. The effects of certain physical and emotional tensions and strains on fluctuations in the level of serum protein-bound iodine. *J. Clin. Endocrin. Metab.*, 20:415.

VON EULER, C., and B. HOLMGREN

1956. The role of hypothalamo-hypophysial connexions in thyroid secretion. *J. Physiol.*, 131:17.

WERNER, S. C.

1961. The thyroid: genetic and psychiatric relations. *Dis. Nerv. Syst.* 22:1.

WHITTAKER, J. O.

1957. Effects of thyroid administration upon avoidance conditioning. *Psych. Reports*, 3:89.

WIENER, J. D., and G. A. LINDEBOOM

1964. The possible occurrence of two inborn errors of iodine metabolism in one patient. *Acta Endocrin.*, 47:385.

WILLIAMS, R. H., and J. L. BAKKE

1962. The thyroid. In R. H. Williams (Ed.), *Textbook of endocrinology*. Philadelphia: W. B. Saunders.

WILSON, W. P., and J. E. JOHNSON

1964. Thyroid hormone and brain function: 1. The EEG in hyperthyroidism with observations on the effect of age, sex, and reserpine in the production of abnormalities. *Electroenceph. Clin. neurophys.*, 16:321.

WILSON, W. P., et al.

1964. Pharmaco-dynamic effects of beta-adrenergic receptor blockade in patients with hyperthyroidism. *J. Clin. Invest.*, 43:1697.

WISWELL, J. G., et al.

1963. Urinary catechol amines and their metabolites in hyperthyroidism and hypothyroidism. *J. Clin. Endocrin.*, 23:1102.

WYNN, J., et al.

1962. Studies of the serum iodine containing components in a family of goiterous individuals. *J. Clin. Endoc. Metab.*, 22:415.

ZILE, M., and H. LARDY

1959. Monoamine oxidase activity in liver of thyroid-fed rats. *Arch. Biochem. and Biophys.*, 82:411.

GENETICS AND CHILD DEVELOPMENT

HANUŠ PAPOUŠEK

BEFORE WE STARTED our behavioral studies in young infants, we sought answers to several basic questions in order to find some reasonable approach. No one of us was a geneticist, and no one was thinking of introducing genetical methods, but the questions had much to do with genetics. Before I start speaking of our results, I will turn back to those questions to make the relationship of our research to genetics understandable. The questions are: only nature or nurture? Genes or higher nervous activity? Individual differences—how and from when to study them?

Only nature or nurture? For a period in its own development, the theory of child development, particularly regarding behavior, suffered from preoccupation with the controversy of "which one: nature or nurture, heredity or environment?" This question came from contemporaneous theories of evolution as a result of certain oversimplification and underestimation of what is human in the human being. Dobzhansky (1957) said that both biologists and sociologists were responsible for that. He explained that evolutionists first had to show what was similar in man and animals before they could pay more attention to what is unique in man and his evolution.

Gradually, it became clear that in the process of phylogenetic development the capacity of using linguistic symbols and developing speech and abstract thinking through them appeared as something substantially new in man and differentiated him sharply from all animals. This capacity enabled man to develop social and cultural activities, to collect experience through generations, and to transmit them to any number of contemporary individuals or to future generations in the form of education. Thus the role of education became in man more important than any other environmental biological determinants.

Analogously, in man the internal integrating mental functions developed into an important determinant themselves. The interesting experimental studies started by Piaget (1926) or Vygotski (1956) in the ontogeny of cognitive processes and in the interaction between speech and thought aided the understanding of origins and development of consciousness in phylogeny. These specific human phenomena affected the evolution of man so profoundly that it cannot be understood without taking them into account.

On the whole, it is now generally accepted that the development of behavior in man is a multifactorial function and that, among the various factors, conscious mental abilities and education, in addition to genetic endowment and biological environmental forces, play the most important role. It may be impossible to measure the relative weight of particular determinants, but we can agree with Anne Anastasi (1958) that for further progress the question "How?" in place of "Which one?" is more important both for theoretical and practical reasons.

Thus we come to the next question of *genes or higher nervous activity*, since these two physiological systems are concerned, and two disciplines of science come mainly into consideration: genetics, including genetical cytology and biochemistry, on one hand and the physiology of brain function on the other hand. Both disciplines and their subjects seem to have been standing at opposite poles. It is not difficult to understand why.

Relatively rigid and stable genetic mechanisms insure the constancy of species above all, and only slow adaptative changes occur during their phylogenetic evolution. But these mechanisms had to bring into the organism a certain antithesis: another type of regulation, much more plastic, capable of quick reactions to countless and often harmful changes in the environment in order to balance homeostasis, to promote health, survival, and reproduction of the carrier of genic endowment during ontogenetic development. It is the nervous system that has developed in higher organisms into a very complicated and specialized structural system governing all kinds of motor or secretory responses—simple functional units that can be compounded into the most complex patterns of behavior.

High in phylogeny, the genetic determination of nervous activity prevailed. Regardless of whether it was due to mutation and selection or due to genetic assimilation of acquired characters, the genetic endowment in fact contains very complex behavioral patterns, which represent a substantial part of nervous activity in insects or lower vertebrates. New abilities of the nervous system then appeared, and the dependence upon genetic endowment became less and less distinct and probable.

Thus, in the ontogenetic-phylogenetic continuum, not only the continuum of relative weight of heredity and environment, but also a continuum of genetic endowment and conscious egocentric mental functions can be demonstrated. In reality, both of these internal factors are interacting and interdependent. Overestimation of one of them caused implacably opposed approaches between genetics and brain function physiology for some period of time. As a matter of fact, these two disciplines are also fully interdependent, and much benefit can be expected from their interaction.

Without any doubt, genetics may bring further discoveries that could solve the key problems of child development, such as the inheritance of acquired characters of population or the ways of supporting, inhibiting, or transforming genetic endowment during ontogeny. But the ranges of genetic research in man are so limited that its main progress is much more probable in infrahuman genetics. The slow reproduction in man, complicated polygenic determination of behavioral traits, and dubious possibilities of breeding are the main limiting factors of human genetics, which leaves perhaps the only advantage in potential registration of some easily detectable abnormalities within large population groups in countries with highly developed preventive health services.

The twin-method, used widely since the thirties and considered valid for study-ing the genotype in man (for example, Lotze, 1937; von Verschuer, 1932; New-man *et al.*, 1937; Luxemburger, 1936; or Siemens, 1932), seems to have exhausted its possibilities, leaving one main, even if not quite new, conclusion that human behavior to a certain extent is determined genetically. Detailed recent studies in identical twins have been focused more upon the degree of genetic background of various mental abilities or large deviations of behavior. Useful information has been collected regarding intelligence or other special abilities (athletic, me-chanical, artistic, or musical) by Wright (1961), Mizuna (1956), Vandenberg (1962), Burt (1957) and others.

Another approach to similar problems has been attempted in longitudinal fol-low-up studies of normal families through several generations, as in the Berkeley project or in Kagan's studies (Kagan and Moss, 1959). These studies show sur-prising constancy of some characters with aging as an end result of complex interaction between all determinants. Here, the analysis of functioning of par-ticular determinants is difficult.

The role of rearing and education has been indirectly redemonstrated in depri-vation situations. Although devised experimentation with deprivation is impossible in children, illustrative data have been obtained by analysis of consequences of various kinds of undesired deprivation. I do not exactly mean the cases of "feral" children where the effects of deprivation cannot be differentiated from those of other factors. But increased attention has been paid to children in institutions, broken families, or unfavorable sociocultural environment, particularly during World War II; and since the basic communications by Spitz (1946), Goldfarb (1943), Bowlby (1951), or Roudinesco and Appel (1950), these problems have been studied by many authors from various aspects.

Interesting results are seen in studies of major behavioral deviations and path-ological aberrations where anomalies in genes caused by structural changes, ab-normal segregation, or nondisjunction during meiosis in the human karyotype have been detected in 23 clinical syndromes with the help of modern tissue culture (Lejeune and Turpin, 1961). Similar detection of genetic traits of differ-ences in normal behavior or in mental abilities would open a new fruitful field for genetic studies in children, but little hope is justified in this direction because of complicated polygenic determination of these functions and effective inter-action of other determinants.

Can neurophysiology or psychology help in this situation? Voronin, when speaking of the reciprocal influence between evolutionary branches of genetics and neurophysiology, explained how profoundly Darwinism influenced Pavlov's concept of the nature of higher nervous activity and Orbeli's studies in the evo-lution of higher nervous activity and, conversely, how their contributions helped to narrow the gap in understanding the interaction between instincts and learning or the development of abstraction, speech, and thinking during phylogeny. These ideas are worth further experimental analysis themselves.

Beside that, we should know more about the basic qualities of central nervous processes and their projection into different mental structures, which is supposed to be the key to inborn individual differences in behavior. Particularly, easier and quantitative detection of such qualities might open new ways for genetical research.

This touches the third question: *individual differences—how and when?* Variability of morphological or functional characters may be of unequal importance for a given organism. Some deviations are limited in their influence to unimportant parts of the organism and are therefore irrelevant for the whole. Other deviations may affect some more important system; thus they can manifest themselves not only in that system but through its functions even in other parts and activities. This must apply above all to brain function coordinating so many activities—hence the traditional association of behavioral and somatic individual differences in so many human typologies, and the tendency to look for basic typological differences in variations of nervous functions.

On the basis of exact physiological experiments in dogs, I. P. Pavlov designed a concept of the nature of so-called higher nervous activity, which included conditioned reflex techniques by which it would be possible to estimate the qualities of basic central nervous processes of excitation or inhibition and to use them even for detecting typological differences. He suggested applying his discoveries to man, and disclosed man's main peculiarities. Conditioning techniques also brought new possibilities in child behavior development. Pavlov himself, and particularly Orbeli, recommended studying the earliest ontogenesis as a necessary step for better understanding of brain function in adults. Although great attention has been paid to these problems in the laboratories of Krasnogorski, Bekhterev, Shtshelovanov, Orbeli, Kasatkin, and others, less effort has been directed to individual differences, except the typological studies of Ivanov-Smolenski (1953) in older children, and those of Volokhov (1953) in infants.

Some authors (for example, Voronin, 1958) do not distinguish individual differences from typological ones as, for example, Young (1947) does. Discussing this problem Janoš (1965), the chief of our laboratory of higher nervous activity, emphasizes that only those various individual differences which concern basic qualities may be considered typological, and that we do not concentrate in our studies only upon typological differences. The question which qualities of the nervous system are basic—that is, typological—has not yet been satisfactorily answered and needs further many-sided research—beyond our resources. In addition, unlike in adults, age differences must be respected in infants because of their rapid development during the first year of life. Therefore, when speaking of individual differences in higher nervous activity, we mean interindividual differences of various kinds, including typological, age, and other differences.

During the first months of life, rapid adaptation of the nervous system to a new environment must occur. Because of methodological difficulties and resulting insufficient knowledge, the importance of this period of development has often been underestimated. Therefore, some unexpected abilities have been demonstrated in newborns during the last years with the introduction of precise and adequate techniques, as in the studies of Kozin (1952), Lipsitt (1963) and other authors.

Ontogenetic studies in experimental pathology show interesting age peculiarities of newborn animals in response to decortication. Instead of irreparable topical changes well defined in adult animals, only transitory and reparable changes appear in newborn (Kogan, 1957; Mysliveček, 1958). In contrast, some other authors (Klosovski, 1960; Křeček, 1962; Lát, 1956) report that disturbances caused by malnutrition, toxic agents, or infections in early ontogenesis can unfavorably influence the whole further development of an animal.

Unevenness of development causing such differences in metabolic demands and in vulnerability of various systems during ontogeny is typical for all species. Thorough knowledge of these age peculiarities is very important for studying forces that shape the phenotype.

There are two more reasons for increased interest in the early development of infant behavior: (1) the structure of mental functions is relatively simple in newborns, and more understandable than later, when it quickly becomes very complicated; and (2) it can be supposed that in newborns the inborn patterns of behavior have been influenced by environmental forces, particularly by social forces or preceding experiences, only to a limited extent.

For all these reasons, studying individual differences of higher nervous activity in human neonates and infants may be a useful approach not only to better understanding of child behavior development, but also from the point of view of the genetical determination of this development.

Now, I wish to draw your attention to some results of our own studies in early development of individual differences in infant behavior, particularly in regard to the problems discussed above.

SUBJECTS

Full-term healthy infants without any evidence of pathology in pregnancy or in delivery are reared in a special unit under relatively standard conditions, with the assistance of mothers and specially trained nurses, who can substitute for mothers if necessary. Babies stay with us approximately six months. Their health state, nutrition, and somatic development are followed by a pediatrician, their mental development and education by a psychologist; both of them are members of our research team. As far as possible with respect to individual demands of infants, we try to keep their life conditions comparable. Up to now, more than 130 infants have been in the care of this unit.

METHODS

Our investigations employ a many-sided approach: working out of conditioned reflexes, differentiation and its reversal as well as analysis of mental development, and behavior with special respect to sleep and waking. In the course of our investigation, we have been faced with a number of methodological problems—optimal methods of rearing, evaluation of existing methods in conditioning, the design of new methods, and the choice of satisfactory criteria for both experimental studies and behavior analysis.

Conditioned eye-blinking technique (Janoš, 1965), head turning toward the source of milk (Papoušek, 1961) or toward the source of optic stimulation (Koch, 1962), analysis of various activities, breathing and EEG during sleep, and transitional stages between sleep and waking (Dittrichová, 1962) have been used for estimating individual differences in higher nervous activity.

THE INFLUENCE OF POSTNATAL AGE

For the purpose of discovering the influence of postnatal age (Janoš, Papoušek, and Dittrichová, 1963), we compared the age of infants at the beginning of individual conditioning processes with the number of experimental sessions necessary for achievement of criteria of conditioning, differentiation, and its reversal.

TABLE 1

RELATIONSHIP BETWEEN AGE AND NUMBER OF SESSIONS TO CRITERION.

Criterion	Reinforcement	n	r	p
Conditioning	Trigeminal	28	− .74	< .001
	Optic	33	− .62	< .001
	Alimentary	55	− .71	< .001
Differentiation	Trigeminal	22	− .50	< .02
	Optic	26	− .43	< .05
	Alimentary	44	− .58	< .001
Reversal of Differentiation	Trigeminal	20	− .22	> .1
	Optic	21	− .15	> .1
	Alimentary	42	− .35	< .02

Table 1 summarizes data obtained in conditioning with three kinds of reinforcement: trigeminal (t—a stream of air) or optic (o—a bright light) in conditioned blinking and alimentary (a—giving milk) in conditioned head turning. The values are shown of the correlation coefficient between age and conditioning (Cr), differentiation (D), and reversal of differentiation (Dr). The relationship between age and rapidity of conditioning is in general highly significant — p < .001. The older the infant, the easier the establishing of conditioned response. The correlation between age and differentiation is also statistically significant.

Despite general agreement that elaboration of conditioned responses is slower in the very young, in both animals and humans, the experimental evidence is not entirely unequivocal. For example, Golubeva and Artemyev (Pantshekova, 1956) believe that conditioned reflexes are elaborated more rapidly in newborn guinea pigs than in adult guinea pigs. Pantshekova (1956) reported that the elaboration occurs at the same rate. Regarding the reversal of differentiation, Troshikhin and Kozlova (1961) found that puppies 1 to 2½ months old reacted more easily than adult dogs. In a whole series of experiments with humans, retardation instead of acceleration in conditioning with age has been reported. The numerous data presented by Razran (1933) to this effect have been amplified by those of Kozin (1952) and Kabanov (1959).

We found that, in the first half year of life, the rapidity of elaboration of conditioned reflexes and differentiation increases with age.

With this conclusion, a further question immediately emerges: comparison and evaluation of these processes may be carred out only within an age range that does not in itself represent significant intrinsic differences. The allowable age span early in postnatal life can be expected to be much narrower than in older groups. How wide is this allowable age span?

The answer to this question has been sought in a comparison of results in infants of different ages. Table 2 presents data obtained by comparison of results in paired groups of different ages at the beginning of conditioning. Designation of applied techniques is the same as in Table 1. It is clear that an age difference of one month or more is almost always associated with a significant difference in the rapidity of appearance of conditioned responses.

TABLE 2

STATISTICAL EVALUATION OF RATES OF CONDITIONING IN VARIOUS AGE GROUPS.

Criterion	Reinforcement	Age difference of groups (days)	Significance of speed difference in conditioning (p)
Conditioning	Alimentary	15.9 (3.5 : 19.4)	> .5
	Alimentary	52.3 (3.5 : 55.8)	< .001
	Alimentary	36.4 (19.4 : 55.8)	< .001
	Trigeminal	43.1 (16.7 : 59.8)	< .001
	Optic	38.0 (18.6 : 56.6)	< .005
	Optic	29.3 (56.6 : 85.9)	> .1

If distortions due to age differences are to be avoided in judging the functional ability of higher nervous activity, then in the early months of life the maximum age range for comparison must not exceed one month. This important ascertainment throws doubt on some previous studies of higher nervous functions in groups of infants differing more than three months in age.

One might expect that the effect of age upon the studied processes would manifest itself not only in the quantitative characteristics, but also in the qualitative patterns, in the dynamic course of these processes. Therefore, in Figure 1 we show the course of conditioning in infants of group A, where conditioning of a blinking reflex was started at 3 weeks of age, and group B, where it was started at 8 weeks. Schematic Vincent curves are employed, representing the average frequency of conditioned responses in individual fifths until achievement of conditioning criteria. While a different form of reinforcement did not substantially change the course of elaboration of conditioned reflexes in each of the two groups, a marked age-dependent factor was apparent; in the older infants in both series we can see an uninterrupted, steeply rising proportion of positive reactions, while in younger infants we find only a slight increase during three or four fifths, whereas in the last one corresponding to the beginning of the third month of life, an abrupt increase in reactions appears, as if a marked qualitative change had occurred.

The dependence of this change upon age is further confirmed by a comparison of the course of elaboration of conditioned reflexes in premature infants (N): while their postnatal age was similar to group B (7 weeks), their corrected age was closer to that of group A (3 weeks). The course of conditioning in these infants was not similar to that of infants of the same postnatal age, but rather to that of infants of the same post-conceptional age.

Analyzing the results obtained with the method of conditioned head turning, we have also observed clear qualitative change in concomitant behavior, the patterns of which are particularly striking when alimentary reinforcement is employed. In the first two months of life, the conditioning stimulus evokes only quantitative changes in general bodily activity and vocalization—nonspecific concomitant behavior. In contrast, by the third month more specific patterns appropriate to the experimental situation appear, for example, various vocal and facial responses reminiscent of adult specific behavior in joy, indecision, uncertainty, displeasure, and so forth. In thirteen infants followed from birth with regard to this aspect, this type of specific concomitant behavior associated with conditioning appeared on the average at 11 weeks of age.

A similar phenomenon can be seen in the development of normal behavior patterns. Analysis of observation samples in twelve infants from 2 to 24 weeks of age

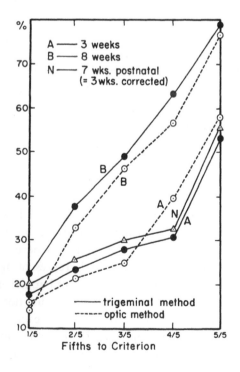

FIGURE 1
The course of conditioning.

shows a striking increase in movements of the hand with a toy and in cooing between 8 and 12 weeks. The increase in both activities is statistically significant.

Thus, from various aspects we have been able to demonstrate marked change in the followed responses at the beginning of the third month of age which in general suggests that at this age there occurs in the development of higher nervous activity a qualitative change the nature of which requires further study.

AGE-DEPENDENT CHARACTERISTICS IN NEWBORNS

Since these characteristics in newborns have repeatedly been a matter of great interest to many authors, it could be perhaps worth mentioning our data on the peculiarities of higher nervous activity in the neonatal period, in other words, in the first four weeks of life (Papoušek, Janoš, and Dittrichová, 1963).

In Figure 2, where conditioning is correlated with age, it can be seen that newborn infants are obviously distinguished by a slow rate of conditioning.

If we compare (Figure 3) Vincent's average curve of attainment of stable food-seeking conditioned reflex in newborns with that observable in 3-month-old infants, we see that it is flatter and even in the last phase does not approach the per cent positive responses of the older infants.

About two thirds of the newborn infants achieve the standard criteria of five consecutive responses within the newborn period. The first conditioned responses, usually weak and with a long latency, appear by the first week of life. Newborns achieved the criteria for the first time at 7, 10, 11, and 12 days. Two infants, after extinction and renewal of the conditioning, completed differentiation by 6 weeks of age, during which a bell signalized milk given from the left, and a buzzer, milk given from the right. In Figure 3, the polygon into which individual curves of conditioning fall is striped.

If the rapidity of conditioning is taken as an index of the strength of the process of excitation, then on the basis of these data the relative weakness of this process can be considered an age-dependent characteristic of newborn infants.

A similar conclusion can be drawn from analysis of waking and sleep states (Figure 4). The waking state characterized by open eyes, frequent general bodily activity, irregular respiration, and vocalization takes up about 10 per cent of the newborn's day (whereas in infants 24 weeks old the waking state occupies 47 per cent—$p < .005$), and is strikingly sporadic: the newborn persists in continuous waking on the average only 6 minutes, as opposed to 64 minutes in the older infants. A considerable portion of the newborn's day is spent in a transitional state between waking and sleep. This state makes up 35 per cent of the day in newborns but falls to 17 per cent in infants 24 weeks old ($p < .01$).

Similarly, in newborn infants, the total time spent in light sleep is longer. On the other hand, the duration of deep sleep, with closed eyes, slow regular breathing, and without generalized activity is the same in newborns as in infants 24 weeks old (approximately 20 per cent), although its course is not so continuous. The frequency of intervals when the eyes remained closed less than ten minutes

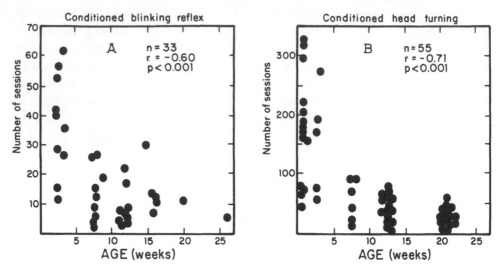

FIGURE 2

Correlation between the rate of conditioning and age.

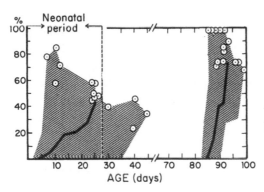

FIGURE 3

The course of conditioning in newborns (n = 14) and three-month-old infants (n = 14).

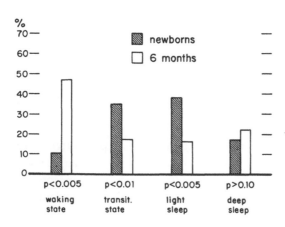

FIGURE 4

Waking and sleeping state in newborns and in six-month-old infants.

falls from 62 per cent in newborns to 44 per cent in the older infants (p < .005). Thus even sleep is sporadic during the first weeks of life and becomes consolidated only later.

The intermittent nature of sleep and waking in the newborn, in regard to experimental conditioning, is probably related to another striking characteristic of this age group—an erratic occasional occurrence of the first positive conditioned responses, and only gradually increasing ability of responding with two, three, or more consecutive conditioned responses (Table 3).

TABLE 3
CONGLOMERATING OF CONDITIONED RESPONSES (HEAD TURNING).

Initial age (days)	n	First 10 Positive Responses		
		Isolated	In groups	
			by 2	3 or more
		%	%	%
3.5	14	61.4	34.2	4.3
85.7	15	36.7	18.7	44.7
		p < .001	p < .05	p < .001

After the first conditioned response, as a rule, temporary inhibition supervenes, obvious even in kymograph records by slow, regular breathing and the disappearance of motor activity. Most of the first conditioned responses are isolated. The ability of conglomerating positive responses grows with age, at different rates in individual infants. In 3-month-old infants most positive responses appear in groups of three to seven.

These observations on the typical inability in the newborns to persist in waking or sleeping state or to carry out several consecutive conditioned responses could indicate in Pavlovian terms that concentration in either excitation or inhibition in newborn infants is not sustained.

Other individual differences that do not depend upon age can be demonstrated in all age groups and by all employed techniques, both in the quantitative and qualitative indicators.

Janoš described several characteristic types of curves in which establishing of conditioned blinking reflex, differentiation, and reversal of differentiation manifested themselves in individual infants. In several cases of double reversal of differentiation, he could demonstrate that the given types of curves were characteristic of individual infants, since in both cases of reversal in the same infants identical types of curves repeated.

In Figure 5, individual differences in conditioning can be seen, as they manifested themselves in conditioned head turning in two groups of infants of the same ages (newborns and 3-month-old infants). In these infants and in another group of 5-month-old infants, we tried to analyze whether a correlation could be found between the rate of conditioning and some other characteristics, such

Newborn
infants (14)

3-month-old
infants (15)

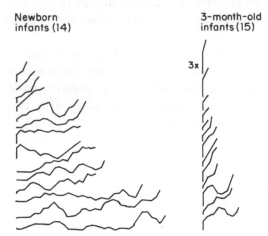

FIGURE 5
Individual differences in conditioning
(head turning).

as birth weight and birth length, rapidity of weight or length increase during the first three months of age, head or chest circumference, and caloric quotient in food-intake during conditioning. The values of correlation quotients are presented in Table 4.

It is obvious that among children of the same age, marked individual differences in higher nervous activity exist that cannot be explained by the variation in age or by the variation in the main anthropometric indicators. There is a significant relationship between the rate of conditioning and the caloric quotient in food-intake in younger groups of infants in the sense that the less the infants drink (or are fed), the more rapid is the conditioning in them. It is quite understandable,

TABLE 4

RELATIONSHIP BETWEEN RATE OF CONDITIONING AND SOME ANTHROPOMETRIC DATA.

	Newborn			*3-month-old infants*			*5-month-old infants*		
	n	*r*	*p*	*n*	*r*	*p*	*n*	*r*	*p*
Birth weight	14	$-.15$	.1	14	$+.01$	.1	14	$+.15$	.1
Weight increase	11	$-.13$	.1	14	$-.006$	.1	12	$-.15$	.1
Birth length	14	$+.11$	.1	12	$+.43$	.1	15	$+.13$	.1
Length increase	11	$-.12$	.1	12	$-.12$	.1	11	$+.15$	.1
Head circumf.	11	$+.03$	.1	14	$+.04$	.1	12	$+.04$	.1
Chest circumf.	11	$-.44$	.1	14	$+.36$	.1	12	$+.38$	.1
Caloric quotient	14	$+.59$	.05	14	$+.63$	.02	12	$+.25$	.1

if we realize the role of milk as a reinforcing stimulus. It is perhaps more interesting that this significant relationship disappears after the third month of life.

We have also analyzed the relationship between the rate of conditioning and the difference in sex, both in conditioned blinking and conditioned head-turning studies, but no significant correlation has been found in infants from birth to 6 months of age.

Further investigation and analysis of more factors will be necessary before we can conclude that we know enough about the nature of observed individual differences or about the role of various factors in determination of the development of child behavior. The data presented in this report can only illustrate the possible ways in this direction; above all, they should illustrate the fact that more needs to be done than has been done up to now.

SUMMARY

Several problems are discussed concerning the relationship to genetics of the author's and his colleagues' studies in individual differences of higher nervous activity in young infants. Contemporary opinion on the main determinants of child development, particularly on the role of the central nervous system and its interaction with the genetic endowment, are the main subjects of discussion. The importance of studying the earliest ontogenetic development of individual differences in higher nervous activity is stressed because of its significance for both child development and genetics.

Experimental data are presented which illustrate the possibilities of this approach, such as the analysis of individual differences in conditioning and in normal behavior with particular reference to sleep and waking in infants. The influence of age is demonstrated in these data together with some peculiarities of higher nervous activity in newborns. An attempt is made to analyze the relationship of observed differences to some anthropometric indicators and to sex differences.

BIBLIOGRAPHY

ANASTASI, A.
    1958. Heredity, environment, and the question "How?" *Psychol. Rev.*, 65:197–208.
BEKHTEREV, V. M., and N. M. SHTSHELOVANOV
    1925. K. obosnovaniiu genetitscheskoi refleksologii [On the basis for a genetic reflexology]. *Sb. Novoe v refleksologii i fiziologii nervnoi sistemy*, 116.
BOWLBY, J.
    1951. *Maternal care and mental health.* Geneva: World Health Organization.
BURT, C.
    1957. Inheritance of mental ability. *Eugen. Rev.*, 49:137–139.
DITTRICHOVÁ, J.
    1962. Nature of sleep in young infants. *J. Appl. Physiol.*, 17:543–546.

DOBZHANSKY, TH.
1957. *Evolution, genetics and man*. New York: John Wiley.

GOLDFARB, W.
1943. Effects of early institutional care on adolescent personality. *J. Exp. Educ.*, 12:106–129.

GOLUBiEVA, J. L., and A. A. ARTEMIEV
1956. (See Pantshekova, E. F., 1956.)

IVANOV-SMOLENSKI, A. G.
1953. Ob izutshenii tipov vysshei nervnoi deiatelnosti zhivotnykh i tsheloveka [Studies of the patterns of higher nervous activity of animals and men]. *Zh. vys. nerv. Deiat.*, 3:36–54.

JANOŠ, O.
1965. Individuální rozdíly ve vyšší nervové činnosti kojenců [Individual differences in the higher nervous system in infants]. Halkova sb. 8 [Halkova Collections, 8]. Prague: SzdN.

JANOŠ, O., H. PAPOUŠEK, and J. DITTRICHOVÁ
1963. Vliv věku na různé projevy VNČ v pruních měsících života [The effect of age upon different functions of the higher nervous system during the first month of life]. *Activ. Nerv. Sup.*, 5:407–410.

KABANOV, A. N.
1959. K voprosu o zakonomernostiakh razvitiia vysshei nervnoi deiatelnosti rebenka [On the question of the regularity of development of the higher nervous activity of the child]. *Trudy 3 nautsh. konf. po vopr. morf., fiziol. i bioch.*, 42–48.

KAGAN, J., and H. A. MOSS
1959. Parental correlates of child's IQ and height: a cross validation of the Berkeley Growth Study results. *Child Develpm.*, 30:325–332.

KASATKIN, N. I.
1948. *Ranniie uslovnyie refleksy v ontogeneze tsheloveka [Early conditioned reflexes in the ontogenesis of man]*. Moscow: Medgiz.

KLOSOVSKI, B. N. (Ed.)
1960. *Problema razvitiia mozga i vliianiia na nego vrednykh faktorov [The problem of the brain and the influence of harmful factors on it]*. Moscow: Medgiz.

KOCH, J.
1962. Die Veränderung des Exzitationsprozesses nach der Nahrungseinnahme und nach dem Schlafe bei Säuglingen im Alter von 5 Monaten. *Z. ärtzl. Forth.*, 55:219–223.

KOGAN, A. B.
1957. *Utsh. zapiski Rostovskogo n. Don. gos. Univerz.* [Published notes on height in Northern Donets State University], 28:109.

KOZIN, N. I.
1952. Vozrastnyie osobennosti korkovoi dinamiki u detei [Age peculiarities of the cortical activity in children]. *Pediatriia*, 34:4, 15–20.

KRASNOGORSKI, N. I.
1958. *Vysshaia nervnaia deiatelnost rebenka [Higher nervous activity of children]*. Leningrad: Medgiz.

KREČEK, J.
1962. *Údobí odstavu a vodní metabolismus [The period of weaning and water metabolism]*. Babákova sb. 28 [Babakova Collection, 28]. Prague: SZdN.

LÁT, J.

1956. O vztahu mezi výživou, přeměnou látek a činností centrálního nervového systému se zvláštním zřetelem k individuálním rozdílům [On the relation between nutrition, the conversion of substances, and the action of the central nervous system, with particular regard to individual differences]. Dissertation, Prague.

LEJEUNE, J., and R. TURPIN

1961. Chromosomal aberrations in man. Amer. J. Human Genet., 13:175–184.

LIPSITT, L. P.

1963. Learning in the first year of life. In L. P. Lipsitt and C. C. Spiker (Eds.), Advances in child development and behavior. New York: Academic Press.

LOTZE, R.

1937. Zwillinge. Einführung in die Zwillingsforschung. Öhringen.

LUXEMBURGER, R.

1936. Die Lehre von den Manifestationschwankungen erblicher Krankheiten. Der Erbarzt.

MIZUNA, T.

1956. Similarity of physique, muscular strength and motor ability in identical twins. Bull. Fac. Educ., Tokyo Univ., 1:190–191.

MYSLIVEČEK, J.

1958. Dynamika vyšší nervové činnosti v dospělosti u normálních krys a u krys, jimž byla po narození vyřazena mozková kůra [The dynamics of higher nervous activity in mature, normal rats and in rats in which the cerebral cortex is removed at birth]. Čs. Fysiol., 7:183–184.

NEWMAN, H. H., R. N. FREEMAN, and K. J. HOLZINGER

1937. Twins: a study of heredity and environment. Chicago: University of Chicago Press.

ORBELI, L. A.

1959. Osobennosti razvitiia vysshei nervnoi deiatelnosti rebenka [Characteristics of the development of the higher nervous activity in children]. Zh. vys. nerv. Deiat., 9:311–318.

PANTSHEKOVA, E. F.

1956. Razvitiie uslovnykh refleksov u belykh krys v ontogeneze [The development of conditioned reflexes in white rats in ontogenesis]. Zh. vys. nerv. Deiat., 6:312–318.

PAPOUŠEK, H.

1961. Conditioned head rotation reflexes in infants in the first months of life. Acta Paediat., 50:565–576.

PAPOUŠEK, H., O. JANOŠ, and J. DITTRICHOVÁ

1963. Razvitiie vysshei nervnoi deiatelnosti rebenka na pervykh mesiatsakh zhizni [The development of higher nervous activity of children in the first month of life]. Materialy 6. nautsh. konf. po vopr. vozrast. morfol., fiziol. i. bioch., Moscow: Izd. APN, 435–436.

PIAGET, J.

1926. The language and thought of the child. London: Kegan Paul.

RAZRAN, G. H. S.

1933. Conditioned responses in children: a behavioral and quantitative critical review of experimental studies. Arch. Psychol., 23:120.

ROUDINESCO, J., and G. APPEL

1950. Les répercussions de la stabulation hospitalière sur le développement psychomoteur des jeunes enfants. Sem. Hôp. Paris., 26:2271–2273.

186 GENETIC DIVERSITY AND HUMAN BEHAVIOR

SIEMENS, H. W.
1932. Die allgemeine Ergebnisse des menschlichen Mehrlingsforschung. *Z. Abstammungslehre*, 2:61.

SPITZ, R. A.
1946. Hospitalism: a follow-up report. *Psychoanal. Study Child.*, 2:113–117.

TROSHIKHIN, V. A., and L. N. KOZLOVA
1961. O stanovlenii i razvitii podvizhnosti i inertnosti nervnykh procesov v ontogeneze [On the formation and development of mobility and the inertness of nervous processes in ontogenesis]. *Zh. vys. nerv. Deiat.*, 11:878–883.

VANDENBERG, S. G.
1962. The hereditary abilities study: hereditary components in a psychological test battery. *Am. J. Human Genet.*, 14:220–237.

VOLOKHOV, A. A.
1953. Tipologitsheskiie osobennosti nervnoi sistemy detei rannego vozrasta [The particular typology of the nervous system of children in early growth]. *Medic. Rabot.*, 16:87.

VON VERSCHUER, O.
1932. Ergebnisse der Zwillingsforschung. *Verhandl. d. Ges. f. Phys. Anthrop.*, 6:1–65.

VORONIN, L. G.
1958. K voprosu o obshtshem i specifitsheskom v filogeneze vysshei nervnoi deiatelnosti [On the question of generality and specificity in the phylogenesis of higher nervous development]. In *Evoliutsiia funkcii nervnoi sistemy* [*Evolution of the function of the nervous system*]. Moscow: Medgiz.

VYGOTSKI, L. S.
1956. *Izbrannyie psikhologitsheskiie issledovanniia* [*Selected psychological analyses*]. Moscow: Izd. APN.

WRIGHT, L.
1961. A study of special abilities in identical twins. *J. Genet. Psychol.*, 99:245–251.

YOUNG, K.
1947. *Personality and problems of adjustment.* New York: Appleton-Century.

CROSS-POPULATION CONSTANCY IN TRAIT PROFILES AND THE STUDY OF THE INHERITANCE OF HUMAN BEHAVIOR VARIABLES

RUTH GUTTMAN

THREE DECADES AGO, Allport and Odbert (1936) compiled a list of 17,953 English terms, each of which specifies in some way a form of human behavior. Such words are invented in accordance with cultural demands; their meanings often vary and some overlap; few, if any, represent "traits freed from the influence of other traits."

The choice of traits for behavior genetic studies represents more than a semantic problem. Psychological variables not only are often poorly defined, they are rarely phenotypes that can be reliably assessed and compared over individuals and groups of different backgrounds.

McClearn (1967) has suggested five criteria for "good" behavioral phenotypes: 1. existence of individual differences in respect to the character studied; 2. adequacy of measurement; 3. minimal gene-environment correlation and interaction; 4. relation of the trait to psychological and chemical evidence; and 5. specificity rather than globality of the trait. The final point has also been stressed by Thompson (1957), Fuller and Thompson (1960), Dilger (1964), Hirsch (1967), and others. These authors discuss the task of identifying and defining the "units," "fragments," or "components" of behavior that can be measured reliably and whose genetic background may eventually be traced.

It is clear that few aspects of human behavior will meet all or even most of the criteria. The problem of gene-environment interaction is particularly complex in the human species, which successfully occupies a large number of physically and culturally variable environments. Furthermore, before one begins to study the variation in any one normal behavior trait—however "unitary" it may appear to be—one must often ask whether this trait actually represents a behavioral or intellectual process common to all mankind.

The possession of specific anatomical, morphological, or physiological characteristics—such as specific tissues, bones, or hemoglobin types—are undoubted characteristics of all men. But species specificity or universality is not always obvious in behavioral measures. Dilger (1964) presented some of the conceptual difficulties concerned with the roles played by genetic and experiential factors in the acquisition and description of species-typical behavior in animals. These difficulties are even greater in studying human traits, especially those observed in terms of performance on different kinds of tests.

"Universals"—regularities in human response which transcend cultural differences—have been discussed by the anthropologists Boas (1911), Wissler (1923), Linton (1936), and others. Kluckhohn (1953) elaborated on "universal categories of culture" and pointed out that "valid cross-cultural comparisons could best proceed from *invariant points of reference* supplied by the biological, psychological and socio-situational 'givens' of human life." Much anthropological research, especially Murdock's large scale cross-cultural survey (1957), has employed cross-cultural methods, usually with respect to cultural characteristics of different groups. Reports on some of the most important studies of this nature have been reprinted in one volume (Moore, 1961), and were recently discussed by Holtzman (1965). Henry and Spiro (1953) listed criteria for the selection of psychological tests for cross-cultural studies. According to the authors, such tests must not be "culture-bound" and the "categories used for their interpretation should be invariant." However, Thompson (1957), among others, questioned whether any test is ever truly "culture-free." The subsequent discussion will propose a definition and technique for ascertaining empirically whether or not a set of variables is "*culture-bound*."

Psychologists generally take for granted the universality of man's behavior traits. To this many add a further assumption of intra-specific uniformity, which ascribes the bulk of individual differences to environmental causation. DuBois (1962) stated that "to many investigators in the field of learning, individual differences among the organisms that are to be the subjects in their studies, are chiefly a nuisance. Groups of subjects . . . are used only to permit averaging results."

Outstanding exceptions to this kind of thinking are the works of Hull and of Thurstone. In his *Essentials of Behavior* (1951), Hull pointed out that the principles presented in his volume were "considered from the point of view of the constancy of natural molar laws," where "both individual and species differences appear in natural molar behavior as *variable values of the constants* involved."

Thurstone (1947) considered one objective of Multiple Factor Analysis to be the discovery of underlying functional unities which produce the test performance studied. These may eventually describe individual differences on the basis of such functions. One requirement the factor model must fulfill, if it is to succeed in discovering basic underlying processes, is that of *invariance*, particularly *configurational invariance* of the same test battery when given to several different populations.

The investigator interested in studying the genetic basis of the variation in behavioral variables is faced with a number of problems which may be either unique to behavioral characters or more complex than in physiological or morphological characters. He must choose for his studies traits that are measurable, repeatable, and comparable from population to population. He must ask himself whether existing variation in the behavior or performance can be compared in some way across different groups, in spite of known variation in geography, cultural background, education, or socioeconomic status.

Such comparison can easily be made for blood groups, for serum proteins, and for some sensory traits such as color blindness. They become more difficult for certain aspects of body proportions which are obviously subject to environmental modifications. Cross-population—and especially cross-cultural—comparison is most difficult and least reliable by means of direct assessment of behavioral tests, even where such measures attempt to be culture-free. Even if exactly the same tests are given to different populations, this does not insure that the same *functions* are involved in the responses of individuals to them. Tests which, for example, in groups of small children measure the perception of new objects, in adults may measure perceptual speed for familiar items. The same may be true for comparison between two populations of adults of equal age, one with a minimum of schooling, the other illiterate.

This paper attempts to illustrate, with examples from the literature and from personal studies, how cross-population invariance or cross-population generality can be established for some trait profiles. It is our hypothesis that, where the intercorrelations between sets of variables of any kind show a constant configuration over culturally or otherwise different populations—especially where variation is known to exist between these populations on any of the measures by themselves—*such configurations are likely to have been produced by processes common to these populations*. While studies on the variation in separate behavioral measures are hampered by problems of interpretation and of environmental effects, configurationally constant trait profiles may suggest universal human phenotypes for genetic, anthropological, or developmental studies.

EXAMPLES OF INVARIANT TRAIT PROFILES

Cross-population generality of intercorrelational patterns has been reported for widely different types of profiles. These include characters with a high genetic component, such as fingerprint patterns, as well as highly complex psychological measures of interpersonal behavior. A few examples will be cited below.

Holt (1952, 1958, 1959) has made extensive studies on the genetics of the number of dermal ridges, which indicate multifactorial inheritance of pattern size. She has calculated mean single ridge counts as well as correlations between ridge counts of the different fingers of each hand in samples of English males and females. In her English samples, adjacent fingers—except for the thumb—are more highly correlated than those farther apart, the size of the correlation decreasing in an orderly fashion from V:IV, to V:III, to V:II, and so on. This correlation pattern, a *simplex*, was the same in all samples, in spite of the fact that mean ridge counts differed for the sexes (Holt, 1958). [A *simplex* (L. Guttman, 1954) is a matrix pattern showing a *simple* order relation for the distances (or similarities or correlations) between *n* things, namely that the *n* things can be represented as points along a line, the plotted distances between the points having a monotone relation to the distance (or similarity) coefficients in the observed matrix.] The pattern was also the same in Mavalwala's (1962) sample of Parsis in India, regard-

TABLE 1

INTERCORRELATIONS AMONG RIDGE COUNTS ON THE DIFFERENT FINGERS OF THE LEFT HAND.
COMPARISON OF HOLT'S ENGLISH WITH MAVALWALA'S INDIAN DATA.
[Adapted from Holt (1959), Tables 3 and 4, and Mavalawala (1962), Tables 1 and 2.]

A. MALES

	English				Indian			
	V	IV	III	II	V	IV	III	II
V	–	.68	.54	.52	–	.66	.50	.48
IV	.68	–	.70	.60	.66	–	.70	.55
III	.54	.70	–	.72	.50	.70	–	.73
II	.52	.60	.72	–	.48	.55	.73	–

B. FEMALES

	V	VI	III	II	V	IV	III	II
V	–	.69	.59	.55	–	.61	.40	.41
IV	.69	–	.72	.60	.61	–	.56	.53
III	.59	.72	–	.68	.40	.56	–	.62
II	.55	.60	.68	–	.41	.53	.62	–

less of significant differences between some of his and Holt's correlation coefficients. Portions of Holt's and Mavalawala's matrices are given in Table 1.

Configurational constancies of intercorrelation patterns across different populations may sometimes indicate developmental relationships of the related measures. One such generality of relationship was suggested for bone lengths in several different animal species (Guttman and Guttman, 1965). Even where the cause of the relationship is unknown, the presence of such constancies is highly suggestive of an underlying similarity of the population samples with respect to the variables forming the pattern.

In the area of mental tests, cross-cultural stability of a factor pattern was reported by Vandenberg (1959) for two sets of twenty tests of identical composition. The first was administered by Thurstone in 1938 to American college students, the second by Vandenberg in 1954 to Chinese students studying at American universities. Vandenberg obtained the same five factors for the American and Chinese data, in spite of the different cultural and educational backgrounds of the two groups of subjects.

A recent reanalysis of responses to sixteen of Thurstone's and Vandenberg's tests with the new nonmetric smallest-space-analysis program (L. Guttman, 1966) has revealed a two-dimensional *radex* configuration which is remarkably similar for the two groups of subjects. [A *radex* (L. Guttman, 1954) is a matrix pattern showing a *doubly ordered* relation in two or more dimensions for the distances (or similarities or correlations) between *n* things; one order relation refers to radial distance from the origin, and the other to the location on the circle (sphere) of a given radius.]

Figures 1 and 2 are the graphic equivalent of portions of Thurstone's and Vandenberg's matrices.

In the figures, two tests are closer as their correlations were larger in the original tables. With minor variations, the same tests appear within the core of the figure; all of these are of an analytic type (starred), achievement tests being in the periphery. The verbal tests (labelled V) are in the upper right-hand region in both groups, while the numerical tests (N) are concentrated on the left side of both figures. The distinction between "analytical," "achievement," "numerical," etc. had been made prior to the analysis on the basis of a classification system described by L. Guttman (1965). The same paper also shows two similar configurations resulting from tests given by Thurstone to eighth-graders and to university students.

Another example of cross-cultural generality was given by Guttman and Guttman (1963) for a pattern of intercorrelations between the scores on six different achievement tests. There a *simplex* structure of identical order was found in forty sub-populations of 15,000 Israeli school children who differed in ethnic background, sex, and type of school, and often differed drastically in their levels of achievement on each subject.

Stafford (1963), in a study of 104 families, found substantial differences between the scores of different family members on a number of mental tests (Table 2). Mothers, on the whole, scored lower than fathers; children scored higher than parents on letter concepts; males higher than females on mental arithmetic. Such group differences are, no doubt, the result of many environmental factors. Yet,

TABLE 2

MEAN SCORES ON FOUR TESTS OF FATHERS, MOTHERS, SONS, AND DAUGHTERS.
[From Stafford's (1962) Table 2, p. 56.]

Test	Fathers N = 104	Mothers N = 104	Sons N = 58	Daughters N = 69
Letter Concepts	5.2	4.1	7.5	7.5
Mental Arithmetic	9.7	5.9	7.7	6.3
Spelling	18.9	20.0	15.2	17.0
English Vocabulary	29.8	27.9	22.1	22.7

in spite of the variations between the sexes and the generations, the tests form a remarkably similar correlational pattern in the four sub-populations. A reanalysis of Stafford's Table 5 shows that the interrelations between the scores on his tests form a simplex pattern of essentially the same configuration in fathers, mothers, sons, and daughters (Table 3). In each group, except possibly in the matrix of the daughters, letter concepts correlates most highly with mental arithmetic, less with spelling, and least with vocabulary. These similarities are apparent despite the small size of the sample, especially of like sexed offspring.

Following our own work, Foa (1964) suggested that certain aspects of interpersonal behavior may be common to different cultures while others change from group to group. In a sample of 633 married couples in Jerusalem, Israel, one subsample originated in Europe, the other in the Middle East. Eight types of inter-

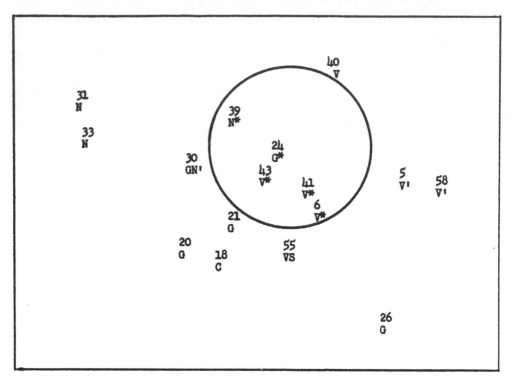

N = numerical   V = verbal   G = geometrical
* = analysis   unstarred = achievement   VS = verbal sound

FIGURE 1

Configuration of sixteen Thurstone mental tests corresponding to correlations obtained from 240 University of Chicago students. Data from Thurstone (1938), Table 2, pp. 110–112. Chart drawn from coordinates calculated by an IBM 7090 program for Guttman-Lingoes Smallest Space Analysis, University of Michigan, December 1964.

Explanation of Symbols

TEST NO.		TEST NO.	
5:	Reading	21:	Form Board
6:	Verbal Classification	24:	Punched Holes
18:	Cubes	26:	Identical Forms
20:	Flaps	30:	Number Code

personal behavior, such as acceptance or rejection of self or of spouse, formed an apparent *circumplex* structure of intercorrelations in both cultural groups, while the actual responses and correlation coefficients varied between the two sub-samples. In a *circumplex*, correlation coefficients first decrease and then increase as one moves away from the diagonal. These configurations are given in Foa's Tables 2 and 3, with his explanations and interpretations in the text.

The final example comes from a recent survey of dental morphology and oral

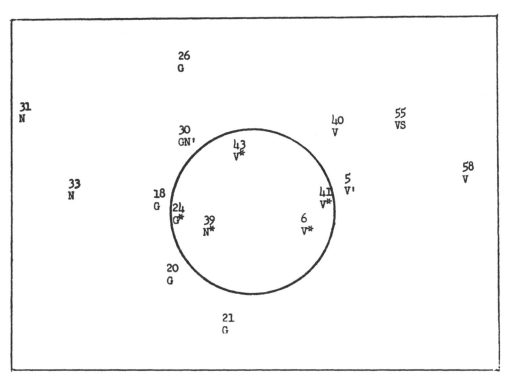

N = numerical   V = verbal   G = geometrical

* = analysis   unstarred = achievement   VS = verbal sound

FIGURE 2

Configuration of sixteen Thurstone mental tests corresponding to correlations obtained from 92 Chinese students in American universities. Data from S. G. Vandenberg (1959), pp. 257–304. Chart drawn from coordinates calculated by an IBM 7090 program for Guttman-Lingoes Smallest Space Analysis, University of Michigan, December 1964.

in Figures 1 and 2

TEST NO.	TEST NO.
31: Addition	41: Verbal Analysis
33: Multiplication	43: Code Words
39: Arithmetical Reasoning	55: Sound Grouping
40: Reasoning	58: Vocabulary

health in several Israeli ethnic groups. This large scale study was conducted by Dr. K. Rosenzweig of the Hebrew University–Hadassah Dental School, with my participation in the non-dental variables. Observations on handclasping, armfolding, and a measure of hand-preference, were made on individuals and families in some Jewish and non-Jewish ethnic groups who live in fifteen Israeli villages (Guttman, Rosenzweig, and Guttman, 1965).

Table 4 summarizes the proportions of persons in the handclasping, armfolding,

TABLE 3

INTERCORRELATIONS AMONG SCORES ON FOUR TESTS OF
FATHERS, MOTHERS, SONS, AND DAUGHTERS.
[From Stafford's (1962) Table 5, p. 61.]

		Fathers				Mothers			
		1	2	3	4	1	2	3	4
Letter Concepts	1.	—	.53	.36	.30	—	.55	.43	.36
Mental Arithmetic	2.	.53	—	.46	.44	.55	—	.53	.46
Spelling	3.	.36	.46	—	.68	.43	.53	—	.55
Vocabulary	4.	.30	.44	.68	—	.36	.46	.55	—
		Sons				Daughters			
Letter Concepts	1.	—	.62	.40	.20	—	.41	.15	.26
Mental Arithmetic	2.	.62	—	.54	.35	.41	—	.29	.31
Spelling	3.	.40	.54	—	.45	.15	.29	—	.50
Vocabulary	4.	.20	.35	.45	—	.26	.31	.50	—

and hand-preference categories. These proportions varied significantly from group to group for handclasping and hand-preference, and the variations were only slight for armfolding. The correlations between the variables also showed large fluctuations from group to group, ranging, for instance, from zero to over 0.9 for the correlation between armfolding and handclasping. Nevertheless, the order between the intercorrelations remained the same for all groups tested. Since three variables can always be arranged in a simplex-like order, the important feature of Table 5 is the *constancy of order* over the the five different ethnic groups. This is discussed in more detail in Guttman and Guttman (1963). We interpret this invariant order to be the result of an underlying process relating to these variables that is common to all the groups (see also Guttman and Guttman, 1965). This point becomes important in considering future studies of handedness, which is well known to have both genetic and environmental components.

TABLE 4

PROPORTIONS OF SOME ISRAELI GROUPS IN HANDCLASPING,
ARMFOLDING, AND HAND-PREFERENCE CATEGORIES.

		Per Cent		
Group	N	Handclasping right over left	Armfolding right over left	Hand-Preference left
Berber	159	60.88	42.43	13.95
Cochin	191	56.88	46.79	17.07
Tripoli	140	68.88	43.32	11.47
Kurdistan	221	56.50	45.01	10.05
Druze	205	43.85	40.35	18.93
All Groups (including others)	1708	60.90	44.51	13.59

TABLE 5

INTERCORRELATIONS AMONG THREE SENSORY VARIABLES
IN SAMPLES FROM FIVE ISRAELI SUBPOPULATIONS.

		Cochin			*Tripoli*			*Kurdistan*		
		1	2	3	1	2	3	1	2	3
Hand-Preference	1.	—	.04	.03	—	.70	.49	—	.81	.40
Armfolding	2.	.04	—	.98	.70	—	.71	.81	—	.66
Handclasping	3.	.03	.98	—	.49	.71	—	.40	.66	—
		Berber			*Druze*			*All Groups*		
		1	2	3	1	2	3	1	2	3
Hand-Preference	1.	—	.98	.57	—	.71	.01	—	.38	.32
Armfolding	2.	.98	—	.57	.71	—	.70	.38	—	.39
Handclasping	3.	.57	.57	—	.01	.70	—	.32	.39	—

DISCUSSION

The examples given above have focused on two kinds of invariance: across cultures and across family members.

Cross-cultural constancy of correlation patterns may afford a definition of the concept "culture-free" and "culture-bound." Let us return to the problem of tests that might be said to measure perception of new objects in children but to measure perceptual speed in adults. How would we actually establish that the tests measure different functions in the two groups? One way would be by giving each group two tests in addition to the one in question: one of objects definitely known and one of items definitely new to each group. On calculating the correlations among all three variables in the two groups, comparison of the two correlation matrices should show that the pattern of correlations of the first variable with the other two is different for children and adults.

Lack of constancy of correlational pattern would, in this case, establish a difference in meaning of the test results for the two groups concerned, which differ in age and experience. The same implication of inconstancy would hold for groups of varying educational or cultural background: a set of tests on variables that do not yield constant cross-cultural patterns can be said to be "culture-bound."

A necessary condition for being "culture-free" is the existence of a cross-cultural generality of pattern. This implies that this concept refers to *sets of variables* rather than to individual ones and that it must be studied as a multivariate problem. However, proving a set of behavioral measures to be culture-free is not sufficient for establishing a genetic basis for these measures and certainly not for analyzing their mode of transmission. The notion of invariance must be extended to include parent-child correlations, using an appropriately extended multivariate model. A model that utilizes both inter-generational and cross-cultural constancy has been presented elsewhere (R. Guttman, 1965).

ACKNOWLEDGMENTS

This paper owes much of its present formulation to the continued advice of Dr. Louis Guttman, who also developed and defined the concepts of *simplex,* *circumplex,* and *radex* as they appear in the text.

BIBLIOGRAPHY

ALLPORT, G. W., and H. S. ODBERT
1936. Trait names: a psycholexical study. *Psychol. Monogr.,* 211.

BOAS, F.
1911. *Mind of primitive man.* New York: Macmillan.

DILGER, W. C.
1964. The interaction between genetic and experiential influences in the development of species-typical behavior. *Amer. Zoologist,* 4:155–160.

DuBOIS, R. H.
1962. The design of correlational studies in training. In R. Glaser (Ed.), *Training research and education.* Pittsburgh: University of Pittsburgh Press.

FOA, U. G.
1964. Cross-cultural similarity and difference in inter-personal behavior. *J. Abnorm. Soc. Psychol.,* 67:517–522.

FULLER, J. S., and W. R. THOMPSON
1960. *Behavior genetics.* New York: John Wiley.

GUTTMAN, L.
1954. A new approach to factor analysis: the radex. In P. F. Lazarsfeld (Ed.), *Mathematical thinking in the social sciences.* Glencoe, Ill.: The Free Press.

1965. The structure of interrelations among intelligence tests. Invitational Conference Testing Problems, Educational Testing Service, Princeton, N.J.

1966. A general nonmetric technique for finding the smallest Euclidian space for a configuration of points. *Psychometrika* (in press).

GUTTMAN, RUTH
1965. A design for the study of the inheritance of normal mental traits. In S. G. Vandenberg (Ed.), *Methods and goals in human behavior genetics.* New York: Academic Press.

GUTTMAN, RUTH, and L. GUTTMAN
1963. Cross-cultural stability of an inter-correlation pattern of abilities: a possible test for a biological basis. *Human Biol.,* 35:53–60.

1965. A new approach to the analysis of growth patterns: the Simplex structure of intercorrelations of measurements. *Growth,* 29:219–232.

GUTTMAN, RUTH, K. A. ROSENZWEIG, and L. GUTTMAN
1965. The retest reliability of the observation of certain traits. *Acta Genet. Statist. Med.,* 15:358–370.

HENRY, J. H., and M. E. SPIRO
1953. Psychological techniques: projective tests in field work. In A. L. Kroeber (Ed.), *Anthropology today.* Chicago: University of Chicago Press.

HIRSCH, J.
1967. Intellectual functioning and the dimensions of human variation. In J. N. Spuhler (Ed.), *Genetic diversity and human behavior*. Chicago: Aldine.

HOLT, S. B.
1952. Genetics of dermal ridges: inheritance of total finger ridge-count. *Ann. Eugen.*, 17:140–161.

1958. Genetics of dermal ridges: the relation between total ridge-count and the variability of counts from finger to finger. *Ann. Human Genet.*, 22:323–339.

1959. The correlation between ridge-counts on different fingers estimated from a population sample. *Ann. Human Genet.*, 23:459–460.

HOLTZMAN, W. H.
1965. Cross-cultural research on personality development. *Human Devel.*, 8:65–86.

HULL, C. L.
1951. *Essentials of Behavior*. New Haven, Conn.: Yale University Press.

KLUCKHOHN, C.
1953. Universal categories of culture. In A. L. Kroeber (Ed.), *Anthropology today*. Chicago: University of Chicago Press.

LINTON, R.
1936. *The study of man*. New York: Appleton-Century-Crofts.

McCLEARN, G. E.
1967. Psychological research and behavioral phenotypes. In J. N. Spuhler (Ed.), *Genetic diversity and human behavior*. Chicago: Aldine.

MAVALWALA, J. D.
1962. Correlations between ridge-counts on all digits of the Parsis of India. *Ann. Human Genet.*, 25:137–138.

MOORE, F. W. (Ed.)
1961. *Readings in cross-cultural methodology*. New Haven: Human Relations Area Files.

MURDOCK, G. P.
1957. World ethnographic sample. *Amer. Anthrop.*, 59:664–687.

STAFFORD, R. E.
1963. An investigation of similarities in parent-child test scores for evidence of hereditary components. *Educ. Testing Serv. Res. Bull.*, RB-63-11. Princeton, N.J.

THOMPSON, W. R.
1957. Traits, factors, and genes. *Eugen. Quart.*, 4:8–16.

THURSTONE, L. L.
1938. Primary mental abilities. *Psychometric Monogr.* No. 1.
1947. *Multiple factor analysis*. Chicago: University of Chicago Press.

VANDENBERG, S. G.
1959. The primary mental abilities of Chinese students: a comparative study of the stability of a factor structure. *Ann. N.Y. Acad. Sci.*, 79:257–304.

WISSLER, C.
1923. *Man and culture*. New York: Thomas Y. Crowell.

IMPLICATIONS OF PRIMATE PALEONTOLOGY
FOR BEHAVIOR

GOTTFRIED KURTH

BEHAVIOR IN A general sense is a complex of different elements. Instinct (innate ability) and intelligence (ability to learn) participate in differing degrees according to age and maturation. Among the primates, intelligence very probably contributed a great deal from the beginning. This applies especially to intragroup behavior or "social" behavior. Generally speaking, genetically fixed changes of behavior will never be fossilized, so any statement about such changes or their possible meaning for evolutionary trends can only be theoretical. We have to make deductions from the morphology of the fossil remains and from the ecological conditions probable for these populations. And we know all too little about this.

For interpretations following recent observations, we must always keep in mind that modern populations generally offer relatively recent annidations ("Einnischung," see Ludwig, 1959, p. 677) in special ecological niches. At best, they may serve as models for the probable complex of conditions of the past. Fortunately, during the last few years, we have obtained so much material by field observations on primates, particularly on the hominoids, that we can take the first steps away from pure theory. Especially important is the insight into the fact that the limits of behavior are considerably larger and the possibilities more manifold than was often expected. On the one hand, this makes it much more difficult to describe the probable behavior of fossil primates, but on the other hand, it now seems much clearer that we should expect fossil primates, especially hominoids at the basal level, to have had fewer clearly distinguishable specializations. On the contrary, we probably have to postulate a markedly open ecotype; and in special degree, this was the fundamental basis for the process of hominisation.

We depend for our statements mainly on morphological observations of the skull and the skeleton. From these and from the probable ecological conditions, we can deduce significant behavioral traits. In this procedure, our ability for making statements is still more restricted by the fact that we have to deduce the possible variation of whole populations or systematic units from fragmentary single parts or, at best, from some few individuals. All this indicates the very hypothetical nature of our conclusions and the caution necessary for our formulations.

In spite of this, it seems possible to work out some aspects which must be considered important during the evolution of the subhuman hominids and which, after passing the "Tier-Mensch Übergangsfeld" (Heberer, 1959, 1965), became increasingly important during the evolution of the human stage until our own period.

Piveteau already explained in his 1961 paper the limitations of our knowledge of the morphology and, thus, the behavior of the probable basal level of primates, especially hominoids. But we are able to say that an essential change must have happened in the first long stage of arboreal life of ancestral primates: the remarkable reduction of the protruding part of the face. Thus the conceiving and observing of the environment must have been slowly transferred from snout and nose—from smelling—to eyes and hands—to seeing and grasping.

During the cerebralization of the mammals, especially the primates and the hominoids, the brain gained several new parts and areas. Phylogenetically older parts became "retracted" or "suppressed" internally—for example, the paleocortex and the limbic system—and are now overlapped by younger, more prominent parts. For man the "basal neocortex" is especially important, as Spatz (1962, 1965) has demonstrated by his excellent research on the brain. This part is also late to develop during human ontogeny. Very probably, these new parts have an higher evolutionary meaning and still retain further evolutionary possibilities.

The increasing use of hands and eyes for conceiving the environment as well as for grasping and bringing in food was, at the same time, an essential prerequisite for the possibility that some of the catarrhine primates could yield some of their traditional arboreal life and could partially or completely change to terrestrial life. In this connection, we must speak very generally because we do not have enough fossil remains and can use only the successful new adaptations or annidations. These were already adapted to the new environment over such a long period that the skeleton demonstrates corresponding morphological changes. Such traits appeared, according to our present knowledge, in the early Miocene. Indeed, it remains uncertain for types like the Proconsulinae whether first known traces of "swinging" brachiation indicate a new evolutionary trend or resulted from an intermediate stage of incomplete adaptation to terrestrial life (Napier, 1959; Napier and Davis, 1959). We shall discuss this question a bit more in connection with the problem of getting erect posture and full bipedalism.

For adaptation to terrestrial life, we have some good examples recently from the Cercopithecoidea—baboons, for example—which have recently been examined extensively. They are still fully quadrupedal with first traces of functional bipedalism, and they seem partially adapted to markedly treeless, rocky habitats. Their hands and feet are already so differentiated as to indicate a relatively long separated evolution. Their diet has a remarkable all-round composition including not only insects, eggs, leaves, roots, and fruits but also the meat of small animals such as young ruminants, which are hunted and eaten only by strong dominant males. In any case, this fact seems to argue that the digestive organs of baboons can utilize animal protein in a sufficient degree. This, I suppose, permits the conclusion that the feeding habits at their basal level were also not too specialized but broad, that is, omnivorous. Washburn and DeVore (1961) in their paper on baboon ecology and human evolution published good observations and photographs of this special kind of feeding habit (see also Kummer and Kurt, 1963).

For our considerations regarding behavior, it seems especially important from the viewpoint of paleontology that the modern strong canines of this type of primate probably developed relatively late and secondarily to this relative size. Fossil remains of a comparable systematic position in earlier stages have mostly smaller canines. This enables us to make a most important conclusion: for terrestrial life in which carnivores were the most dangerous enemies, the defense of the group depended to a high degree on social cooperation. Such marked intragroup cooperation already in the early basal level appears to us as a significant trait of behavior. It seems particularly essential because we must consider this kind of behavior as a decisive prerequisite for the whole process of hominisation. But before we consider this, we shall have a short look at some fossil remains.

First, we have to mention the sequence: *Propliopithecus, Pliopithecus, Epipliopithecus*. They cover the span from the second half of the Miocene until the middle of the Pliocene and are known from North Africa (Egypt) and Europe (Moravia, Lower Austria). They have been best studied by Zapfe of Vienna. For this group it seems important that their fore- and hindlimbs had about the same length until the middle of the Pliocene. The fissure of Neudorf-on-the-March offered excellent skeletal remains for this conclusion. Their proportions (Zapfe, 1958, 1961, 1963) indicate a quadrupedal terrestrial life in a remarkably rocky habitat. This seems particularly meaningful if we regard, with Zapfe, this type as probably ancestral to the recent hylobatids, that is, our most extreme swinging brachiators. Hence their specializations should be considered relatively new. This indicates again that, according to such observations, the main area of hominoids should have been more in open districts or on the border of larger forests over a long time. (See also Bergounioux and Crouzel, 1965; Pilbeam and Simons, 1965; and Piveteau 1957, 1961.)

We can presuppose similar conditions for the much earlier Proconsulinae as well as *Limnopithecus* from the same area and period. Both we know only from East Africa. They can be dated in the first half of the Miocene. For *Limnopithecus*, it seems sufficient to mention in this connection that this type was probably quadrupedal and led a terrestrial life (Ferembach, 1958). There are no clear trends in the directions of special modern adaptations, while the Proconsulinae offer much more in evolutionary meaning. On the one hand, we have a relatively large number of individuals from this group; their remains permit a good reconstruction of skull and skeleton. On the other hand, this type is very interesting because the adults ranged in size from that of recent chimpanzee to that of recent gorilla. However this does not permit the conclusion that their habitat was similar or fully comparable with recent times.

At present, we can only say that they lived mainly in open areas and led a terrestrial life. Napier (1959) and Napier and Davis (1959) have already said that the foot bones—calcaneus and talus—seem to demonstrate first trends in the direction of a beginning brachiation, but the limb proportions offer no remarkable differentiation in the direction of a special "swinging" brachiation, in other words, longer arms and relatively shortened legs.

With this, we reach the important problem of getting erect posture. This process, decisive for hominisation, very probably depended on life in mainly open areas. Thus selection pressure favored again and again the offspring of individuals who could change functionally from a pronograde quadrupedal posture to a bipedal posture for a longer time. Naturally this never excluded climbing trees, and the functional bipedalism may even have been favored by the fact that this type had to climb trees frequently. But the mixed posture with a steady change between quadrupedal walking and functional bipedalism seems to have been decisive for the open ecotype of the basal level.

A very meaningful example of this evolutionary trend is the fossil remains of *Oreopithecus* (Hürzeler, 1958). This important type is dated about ten to twelve million years ago at approximately the transitional stage between the Miocene and the Pliocene. There is still no full agreement on the systematic position of these remains. But in contradiction to the old view that this type should be placed in the Cercopithecoidea, it now appears more probable that *Oreopithecus* will be placed with better justification in the hominoids and particularly in the hominids. During the long discussions over this problem, very different ways of life were attributed to the Oreopithecinae. One discussed possibility was a predominantly arboreal life in a swampy environment (see, for example, Schultz, 1960). But now new investigations by Kummer in 1965 offer new aspects for our problem.

According to Kummer, the distal ends of the femora demonstrate such a clearly angular position that this could only be explained in the sense of a genu valgum. Whereas in orangs and gibbons the collum femoris is steeply erect, its position in *Oreopithecus* is much more horizontal. Again this combined with the position of the distal ends of the femora indicates a high degree of bipedalism. Kälin (personal communication) completes this interpretation with his studies which show that the *Oreopithecus* calcaneus and talus have a structure similar to *Proconsul*. This could indicate a possible trend in the direction of brachiation (Napier, 1959; Napier and Davis, 1959). This is not contradictory to Kummer's view because the process leading to erect posture could have been supported by steady efforts to climb about in trees. But the strength of the femur in *Oreopithecus* seems to indicate that brachiation as a special adaptation never was a prerequisite for hominid bipedalism.

In this connection we have to mention some observations on recent primate populations from Japan, which underline the fact that open ecotypes on a relatively low level with widest possibilities for adaptation and with large ability to learn when a new ecological niche is offered, in this case one intentionally stimulated by the Japanese colleagues. An isolated island population of some hundred macaques were offered additional food; they "learned" to clean and wash this food, and, during the transport of this food to the sea for washing, to walk in an erect posture for about fifty meters.

Considering the new mode of locomotion and the broadness of possibilities used only under special conditions, this process offers some especially interesting points. The new possibilities were explored by young individuals, mostly females. Some of the adults did not take over the new habits, but all individuals born after the opening of this new way did. This offers significant insights for the human phase of hominisation in which tradition became increasingly meaningful within "societies" as an additional new way for transferring information. In this, one has to remember the tendency toward conservatism demonstrated in cultural history by the long preservation of some techniques, tool-types, and ways of life over very long time periods. It can be seen in the discrepancy between the relatively slower change of the transmitted "culture" compared with the clearer change of the morphology of the transmitting populations. This may be interpreted in the sense that the weight and influence of the aged members within groups or "societies" act or may act more for conservatism than for innovation or change.

Thus recent field research, morphological observations, and preliminary deductions from our fossil material offer some consequences for our purpose. They could be significant for our conceptions about the behavior of populations at the basal hominoid level upon reaching the open field and the process of hominisation. On the one side, we can say in spite of the poor fossil remains recovered so far that the time necessary for this process must be calculated as starting about twenty million years ago. On the other side, it is probable that the hominoids altogether lived a more terrestrial than arboreal life during a very large portion of this time. This seems to be indicated by the fact that we find the Hylobatidae as well as the Pongidae developing the first clear traits of a more specialized skeletal adaptation for brachiation only relatively late. This raises the question of whether their recent behavior contains some indications for the supposition that they had a broader adaptation in their remarkably narrow ecological niche in the past than today.

Concerning this, we have the interesting observations of Barbara Harrison, the zoologist so well known for her efforts to preserve Pongo in Northern Borneo. She buys captive young orangs and keeps them in a large parklike area. According to information from newspapers, these half-tamed orangs go into some natural caves found in this park to sleep. This is very interesting since this park offers many trees as sleeping places for the "arboreal" ape. On the other hand, this observation offers a probable explanation for the fact that we get so many orang teeth from glacial levels in caves of the Asiatic continent corresponding to similar finds by Dubois in caves of Sumatra. These facts are interesting since, among the apes, *Pongo* has the most extreme brachiator proportions and leads a particularly arboreal life in recent habitats.

Among the African apes, the gorilla appears to be less interesting for our discussion. His behavior seems partially influenced by the fact that his relatively heavy body required a secondary return to a terrestrial life. Recent observations of chimpanzees in the wild by Jane Goodall may indicate an earlier level with broader adaptation in hominoids. In some troops, chimpanzees were seen hunting antelopes or monkeys cooperatively, the dominant male sharing the meat of the prey with favored members of the troop. Although group-hunting and food-sharing have been reported for only a few troops, these observations may, with caution, be taken to indicate that a mixed diet including animal protein was essential for early hominoids prior to the special vegetarian feeding adaptation of surviving apes.

With this, we have reached a certain basis for our considerations, probably significant for our theories concerning the possible behavior of those populations which entered the savanna and the evolutionary trend toward hominisation at the same time. In this connection, we need not discuss similarities of expression and gesture demonstrated by pongids and hominids during special situations, such as aggression, fury, fright, pleasure, and pain. More important is the social behavior of groups which might indicate essential consequences for the social behavior of modern human populations. For this purpose, observations on recent hylobatids and recent apes may be relevant only within limits. First of all, we have to exclude all groups which lead a well developed arboreal life since this seems to be a particular and relatively new annidation reached independently from the process of hominisation. Again, unfortunately, we must depend on purely theoretical considerations. With the exception of the side-branch *Oreopithecus*, we still have no Hominidae fossil remains from the Miocene or the Pliocene, the long subhuman stage of hominisation. Only since the early Pleistocene do we have fossil remains of "human" hominids together with tools and thus a fully developed human ability.

First we are again involved with some morphological considerations. During the slow and long transition from functional to permanent bipedalism with fully erect posture, the hands became free for new purposes to an increasing degree. During this process, the skull showed further reduction of the snout. This can be seen from the new reconstruction of the *Proconsul* skull of the early Miocene by Davis and Napier (1963). We must remember that absolutely larger mandibles and maxillae had relatively larger canines. In contrast, it is significant that *Oreopithecus*, besides generally lacking diastemas, has small canines, not only absolutely but also relatively. This leads to the important conclusion that, during the extended subhuman stage of hominisation, strong teeth, particularly canines, became less necessary as a means of defense.

Insofar as we are able to reconstruct the possible habitats of such hominid populations, we must assume that the open environment which was conducive to the development of erect posture also included carnivorous predators, particularly felids, as potential and dangerous enemies. The freeing of the hands has very probably helped in the use of naturally available tools. But we cannot really expect with any certainty that such tools could have been of much use as a means of defense against predators. At the same time, we cannot expect that an initial cerebralization could have offered the subhuman hominids any remarkable assistance through intelligence. Therefore, we must consider again that the close social cooperation within hominid populations provided the main decisive element for mutual protection from predators. (See also Pilbeam and Simons, 1965.)

This social cooperation can be exemplified by recent observations of baboons and chimpanzees. In both, the troop includes, on the average, a large number of individuals of both sexes and of every age, the range of ages seldom exceeding the duration of one generation. There is a clear order of precedence with several dominant males. The subdominants defer to the single or few ruling males. According to the calculations from recently published field observations, the number of individuals in one troop seems never to exceed a size that makes it impossible for each single troop member to know all others as well as to have real interpersonal contacts with them. The numbers range from a few to some hundred individuals; for example, Kummer and Kurt (1963) estimate from 12 to 750. But the large numbers are a short-time effect of many neighboring troops simultaneously leaving their sleeping places in the morning—for example, Hamadryas baboons—and these numbers are not relevant for typical troop size. In fact, such aggregations often include a number of small units consisting of single males with one to nine females. Generally speaking, it seems possible to conclude that, during longtime selection pressure, the advantage of mutual protection of all members of one troop lowered intragroup tensions to a possible minimum.

This trend seems to have been increasingly effective for the subhuman hominids because they survived, according to our present fragmentary knowledge, the long stage of hominisation without marked specializations until they reached the special "human" ecological niche and the particularly hominine—human—abilities. With all caution, we can perhaps offer the statement that an accentuated social behavior within small groups—from the neonatus to the maturus—was selected as necessary for survival. This trend must have increased and become more important for the hominids—compared with hominoids—because they missed all specialized possibilities for defense during the long subhuman stage of hominisation while their new "human" abilities were still undeveloped or in their initial stage.

Moreover, it seems probable that during the long stage of hominisation an additional loss occurred in the reserve of relatively schematized behavioral traits that limit the effect of intraspecific tensions in other animals. As essential proof for this loss, we point out that the earliest fossil remains of human hominids demonstrate clear signs of cannibalism. This indicates that their innate behavior could no longer prevent them from killing and eating members of their own species. Assuming the very mixed diet of the hominid, we cannot presuppose that the smell of blood stimulated the eating of hominids killed during combat as sometimes happens among some of the Canidae, for example, wolves. Therefore we have to look for another explanation for this peculiar behavior of the "human" hominids.

As I see it, it seems probable that the stimulating factor could be the fact that small groups had to survive and live as open ecotypes in units which were quantitatively very limited in spite of their worldwide distribution. Perhaps an indication of this is the fact that many tribe names of *Homo sapiens sapiens* mean nothing else than "man." Members of neighboring tribes with very similar morphology can indeed enter the tribe with this special name; but it is impossible to call them as members of a "foreign" tribe with their own special denomination which is strictly reserved for their own tribe, for example, *Khoi* or *San* for the Hottentots or Bushmen of South Africa or *Iki* for an American Indian band of California. Additionally, we must mention the often very different languages of tribes in close proximity which underline a striking trend of the hominines, namely, to split up and to separate into very small units.

But, under phylogenetical aspects, this indicates only that we have to regard an accentuated social intragroup behavior within limited units as essential or even decisive for the hominids, not only during the whole process of hominisation but also during the human stage. Indeed this raises the question of how this kind of behavior will be fixed genetically and the phylogenetical meaning of changes in the mutation rate. Once more, we have to remember that during the subhuman stage of hominisation, the achievement of full bipedalism with the freeing of the hands was not combined with an observable cerebralization at the same time or to the same degree. Any statements in this direction are to be based only upon the estimated mass of the brain or the capacity in relation to the probable mass or weight of the whole body. But we have no evidence for possible changes in the fine-structure of the brain and its performance. Recent observations of apes prove that tool using occurs frequently and even toolmaking is evident. The brain weight of recent apes with such abilities clearly overlaps the variation of the known earliest "human" hominids.

In this connection the problem remains if changes in complex "behavior" thereby could have been decisive for the particularly human abilities (Heberer, 1959, 1965; Koenigswald, 1965; Tobias, 1965) and how we may define this quality. Generally, we notice that primates, especially the hominoids, show an increased curiosity and a corresponding tendency to play. This should not be considered an effect of the absolutely as well as relatively extended stage of childhood and youth. But this alone may never have been sufficient to open the new ecological niche for the "human" ability so significant for the "human" hominids. Perhaps we could say cautiously that the new quality, "perseverance," must have been the decisive and necessary additive for the completion of the new behavioral mode insofar as we may generally define special single factors of such very complex processes. Regarding this, absolutely nothing can be said about the highly probable multi-factorial genetical basis of so-called single "factors."

This new quality, perseverance, may have made possible the transition to intentional working by combining some qualities already existing in primates with the increased degree of intelligence in the hominoids. During the last part of this stage, there was no longer just tool using but an intentional toolmaking, the important step. Since toolmaking has been proved for apes also, we must consider how we can separate the hominids from the pongids from this viewpoint. Kälin (personal communication) developed a useful solution during a discussion with Buytendijk which seems acceptable for all sides; according to both, we should select as the significant criterion of "a fossil man" the fact that those individuals not only made tools—Geräte!—but also made the same type of tool again and again.

With this, we come to a further significant or decisive criterion for the "human ability." For the human hominids, the well-known biological transmission of traits is complemented more and more by the possibility of transferring observations and knowledge independently of gene transport during the sequence of generations only by personal contact between individuals and groups of individuals. We should regard this additive way not only as a simple consequence of extended childhood and youth but also as a strengthening of intragroup influence particularly for the more aged individuals.

During the long "scouting" period of the new ecological niche by human hominids, this opened a new trend or direction for selection pressure. During this period and probably very slowly, old age with its tested observations and knowledge could become increasingly valuable in the new field of human ability even after the reproductive age. According to the fossil evidence, this new trend became really effective only slowly and relatively late, similar to the slow perfecting of cerebralization. The same direction seems to be indicated by the slow technical improvement and perfection not only of tool types but also of their use in work. All this picks up in tempo only after the appearance of our own species and particularly our own subspecies, *Homo sapiens sapiens*. The tool types attributed to the australopithecines remained very similar for more than one million years.

In spite of this, we may generally say that the evolution of hominids seems to have been directed from the beginning toward relative independence from less flexible and therefore schematized innate modes of behavior to the intraspecific social behavior within groups. Regarding this problem, we have already stated that modes of behavior in the primates and particularly in the hominoids must have included a high portion of plasticity from the beginning and therefore represented a complex or combination of instinct and intelligence. For the hominids, this "plastic" component as the basis of intelligence became remarkably important. But, on the other hand, this became possible only because those portions of innate behavior that were important for intraspecific behavior had decreased, as may be shown by the early practice of cannibalism. But this seems mainly related to the social or intragroup behavior, whereas the innate or instinctive modes of behavior for propagation and the first stage after birth seem to have remained more or less unweakened.

This differential distribution of losses indicates very essential aspects of the hominisation process. For the transfer of observations, experiences, and knowledge of personal contact between individuals, a few symbols seem to suffice within the group. For our purpose, we may leave out the type of symbol. Modern ethnological research has sufficiently established that a socially and technically highly organised way of life in the postglacial period (Mesolithic or Neolithic) seems to have been satisfied with languages with a vocabulary never surpassing a few hundred words. Therefore we need not discuss the problem as long as the structure of the mandibles in (pre)hominines and early (eu)hominines may indicate a well-developed ability for talking.

The evidence for cultural tradition in tool types similar over a long period, at present known only for hominines, underlines the necessity to presuppose a minimum of symbols for transferring information. These symbols must have been adaptable to the popular mind but, at the same time, must have needed a comparable degree of intelligence for all participating group members. In the same way, the importance of the group must have further increased so that the traditional dominance of a few members may have slowly shifted more to the whole group as a social community. But all this seems to depend to an increasing degree on. a sufficient flexibility. That the required flexibility was probably provided by the groups very early is supported by the fact that special symbols for the transfer of observations and knowledge were known only to the small unit whose members stood in permanent exchange. The importance of such units may have been enhanced on the one side by the fact that the extension of childhood and youth strengthened the biological unit for propagation, the family. On the other side, the increasing significance of intelligence for influencing the group as a unit for the transfer of knowledge probably heightened the value of members too old for reproduction. The age distribution of the hominids soon included more than one and a half generations and reached at least nearly two generations, or a span of about sixty years. This strengthened the weight of groups as social units in a larger sense, and, as a consequence, more and more learned traits became available for social behavior within the group.

In this connection, we must again refer to the outstanding "human" characteristic, cannibalism. Its appearance underlines clearly enough the loss of innate standards maintained by instinctive, schematized behavior. Indeed, it is unnecessary to emphasize that ethnologically known cannibalism with ritual or magic cannot be a model for the kind of cannibalism verified archeologically for the early stages of human hominids by undoubted traces on the fossil remains. On the other hand, we may presuppose that a mutual social behavior of group members must have been provided among the australopithecines through some—*cum grano salis*—social tradition. Considering the worldwide distribution of this open ecotype with very limited numbers, they must have been remarkably group-directed.

But in spite of some convergencies due to similarities of life conditions, there must have been many different solutions for the usual social needs. This seems to be confirmed by ethnological observations on recent populations that practice widely diverse arrangements of group life. In this, we again notice that the group as a social unit has a clear preponderance over its individual members. This becomes evident from the fact that mothers during very bad times kill their newborn and use them for food for the other offspring (Australia). On the other side of the age pyramid, we have examples from some unsuitable habitats where the aged voluntarily seek their own death when they are no longer able to get their own food and thus, by depending on the efforts of others, endanger the survival of the group (Eskimo) or where they may agree to be killed by a son (Indians of Tierra del Fuego). In this connection we may mention human sacrifices practiced as a religious ceremony for the safety of larger communities.

Regarding the whole problem, we may say in general that the behavior of the group members to each other follows certain measures that probably sprang from a mutual *do ut des* relation. It appears particularly essential that the growing members of the group can observe and learn such "social" intragroup behavior unconsciously during their early childhood before they are officially introduced to the "ritualized rules." In larger units, these rules will be transferred by moral/ ethical, metaphysical, and political/ideological systems of values which direct the social life of communities. But their persuasive power depends on whether or not, and to what degree, they can be practically confirmed by the everyday experience of single members within smaller groups. To accomplish this, the group should never be so large as to exclude personal knowledge of and personal contact with all members of the group.

With this, we arrive at numbers that may roughly correspond to the size of small "natural" units made up of a large family with at least two generations. We may presuppose this size to be the usual basis for the whole human stage of hominisation until the end of the Pleistocene. We may reflect on the probability that estimations of the size of the whole world population before about 12 to 15,000 years ago are generally only a few million and estimations of under two million are the most probable (Kurth, 1965a, 1965b).

What consequences result from this for our considerations of the intragroup behavior of human populations? The generally small number *in toto* and the very low population density together with the worldwide distribution could probably never compensate for the loss of innate barriers in the intraspecific behavior. It seems evident from the human cannibalism during the whole Pleistocene that the hominid trend for increasing independence from instinct in social behavior, once adopted, retained its validity. Therefore, it seems hardly probable that we could detect some additional genetically fixed modes of behavior. On the contrary, the increasing influence of intelligence for the regulation of social behavior within the extended family groups may have still strengthened the old trend for relative isolation.

The biological effect of limitation of possible breeding contacts can be seen again and again in the morphologically and metrically determinable differences between regionally representative series. This is not surprising if one takes into consideration the growing influence of intelligence in all cultural affairs, and we may include in this language and social organization. The considerable differences among neighboring, so-called primitive peoples (Naturvölker!) are well known. This is supported over longer distances by the postglacial differentiation of regional races, populations with more frequent gene-exchange within the race than with those of other regional races. This differentiation has continued through the many millenia of postglacial time during which the proliferating world population doubled in ever shorter intervals. But the rapid increase in the rate of population growth is only the effect of the modern technical civilization and hygiene, especially medicine. The limit of about a hundred million for the whole world population was surpassed about 2,500 years ago, the limit of one billion being surpassed in just the last century.

From this, several important consequences come forth. The small group, selected and established as the unit for propagation and social cooperation during millions of years, was very probably never exposed in post-Pleistocene times to a selection pressure that could have caused any remarkable change in the genetical basis for intraspecific "social" behavior. Until the second half of our millenium, we must still estimate that the doubling intervals were rather long for the world population. These doubling intervals probably decreased from about 2,500 years to about 800 years during the period of about 12,000 b.p. to about 300 b.p. Thus enough time remained to adapt the learned measures for social behavior, transmitted as a kind of longtime "absolute" value system, so slowly that the individuals generally would never have felt a recognized deprivation in their traditional respect for such systems. Also, in the classical civilizations, the individually surveyable small units retained their weight for experiencing and confirming the meaning of the transferred norms for social behavior through everyday practice. And we know still later, in extended areas under the cultural influence of Islam or in the Chinese Empire, the large family was still dominant. In part, this was reinforced by the habit of craftsmen of the same profession as well as tribesmen having their living quarters close together.

Only since the last few centuries has the life expectancy of the newborn increased rapidly by means of medicine and hygiene. During the same time, the technical civilization has offered such variety of work for the support of a family that many more adults were able to marry and this at a considerably earlier age than before.

So started the modern population explosion, and soon, in some regions, it rose to high densities of millions of people. Within those areas, small units almost completely lost their former meaning. On the one hand, the large family containing at least two generations disintegrated and thus the unit for propagation and social education disintegrated. On the other hand, the extended groups in which each member interacted with all others and which served as units for social education beyond the family also disappeared very rapidly. These extended units can never be replaced by the neighborhood because such social duties as joint work, so effective in village life and partially in small town life, are missing in larger groupings. The larger professional, ideological, or confessional groups—trade unions, political parties, or religions—are no more able to fill the place of the small units than are the neighborhoods. They are too large and, for individuals, no longer offer personal contacts.

Under such conditions, the fundamental measures formed by the traditional combination of innate and learned modes of behavior and confirmed by daily experience are missing; these measures support a biologically meaningful life for individuals in mass societies. Life in small units seems particularly worthwhile for man in that single members may, in spite of different status positions, confirm their personal value as members of their group. In the same sense, "asocial" behavior will be punished by exclusion.

Perhaps recent observations in industry may indicate the importance of the self-estimation of the individual in the knowledge that he has a necessary role in the efficiency of the group. For this, we have some informative inquiries by the sociologist Kellner (1963) who examined factories with a high percentage of ill workers in the German Federal Republic. These factories had the most modern equipment, a five-day week, very good salaries, and additional social benefits. Yet personal conversations at the factory revealed that workers were dissatisfied with the work plans and the oft-repeated interventions from above. After the management agreed that the worker could have more independence in personally planning his work and more responsibility, not only did the output increase but also the number of ill workers decreased very rapidly. This seems to be confirmed by comparable results in different factories. Moreover, it is meaningful that the workers were much less interested in the controlling influence of their trade union in factory management and economy (Mitbestimmung) because, as they said, they could never know the larger problems well enough; but they were very interested in their own small responsibility within their own domain.

Within the modern industrial society, it is possible that essential traits of former behavioral modes within small units with face-to-face contact may still be active. For the satisfaction and self-respect of single individuals, a confirmed role is needed within a small unit. Such a role, centered about the workshop, may replace the more complex social unit previously so important for self-respect. In this connection, it appears that for the individual, not only are the "learned" modes of behavior meaningful but also at least as meaningful is his group position, which can be satisfactorily felt and understood only emotionally. Again this may support the assumption that the complex of factors directing the whole behavior of human individuals has a very broad basis. This makes it still more difficult to analyze theoretically the obviously multifactorial gene basis of our behavior.

Regarding our modern mass society, we may perhaps deduce from these considerations and observations that there are some possibilities of activating traditional modes of behavior within limits, in spite of the destruction of the former small social units in mass populations. As one way, we probably have to improve and confirm the individual's self-respect as a member of primary social groups.

This is particularly relevant for the so-called under-developed peoples. They still live with a high degree of interpersonal contacts and cooperation, some of which are supported by means of rituals. We have to look for the greatest possible degree of preservation of such contacts during the transfer of these people to modern western modes of society and economy. The experiment needs special understanding and much empathy because, as an effect of our "human" ability to change our environment, some factors will be altered very rapidly and drastically, and this trend may still increase. Instead of the traditionally low rate of population growth with very limited population pressure, new possibilities in a modern environment with science and technology result in population explosions. While all changes happen very rapidly and since the population pressure is at present (and probably for at least the next two generations) enormously high, we have no time for slow adaptations; we are particularly unable to predict any possible or probable trend in selection. At the same time, we must take into consideration the mainly emotional basis for the acting and reacting of human individuals or groups and particularly of mass societies.

Finally we should briefly discuss two phylogenetic aspects of the problem that could be meaningful for the behavior of recent human populations. Insofar as we know living and feeding habits during the human phase of hominisation, we find that, besides gathering, hunting was especially important. With such irregular amounts of food on the average, it must have been crucial for the hunter that his body could release large, if short, bursts of energy, even though his state of nutrition was bad after a long period of hunger. This would be particularly true if weapons and other means of getting game were inefficient and the game had to be worn out through long chases.

In any case, we may expect that selection pressure acted through millions of years in the same direction, that is, to free by appropriate stimuli all available funds of energy in the human body. This physiological effect could have been developed as an easy response of existing metabolic capacity. During the long life in small bands with a very low population density, these outbursts of energy most likely had no dangerous consequences within the "society." This has probably changed ever since permanent settlement, intentional food production, and increased population numbers have heightened the possibilities for tension.

Perhaps a cautious and theoretical interpretation can be made for the long past that group-dancing at night, especially of youth, as well as corresponding activities during tribal meetings and festivals served as means of skimming off excessive or easily stimulated energies through repeated vigorous exertions. In our modern technological civilization, the surplus energies of the human body are skimmed off or channeled less and less; in contrast, modern man steadily receives stimuli freeing energy in his "older" metabolism. Unexpected mass or mob reactions are possibly the simple expression of an "old" regulatory mode of our metabolism, selected through millions of years for quite different demands as well as for living in small bands with low population densities. Perhaps this gives us some insight into recent problems of human behavior and of mass reaction.

The second area of our problem is the recent change of age distribution within our populations. During hominisation, the middle-aged and the aged may have obtained a special importance for the survival of the group by transmitting their rich life experience. This may be correct even though the observations of our Japanese colleagues (see pages 202–203) demonstrate that among macaques it is the young individuals that explore new niches or possibilities first. But for consistency of groups or societies among hominids, the aged were, very probably, always the most important.

As far as we know the approximate age distribution of fossil or subfossil small populations, the percentage of people over fifty was always very low. Therefore it may not be erroneous to suppose that the few individuals surviving to senility were physiologically young or very vital considering their age in years. Within our civilization, the percentage of people over sixty has increased considerably and, in the industrialized countries, has already reached nearly one third of all adults. Among these, we find many who reach this age only with the help of modern medicine. Many of our elderly people are really senile and, on the average, are much less vital than the very few aged of earlier periods.

Within recent populations of our civilized societies, their influence and importance is much higher due to their actual numbers; this may be multiplied during the long phases of testing the retarding effect of this age class with their tendency for conservation and security. Additionally, the acceleration of technical development gives the youngest generation a quite different basis for their experience than was possible for the middle-aged, not to mention the really aged. It becomes more and more difficult for the three age classes to find a common basis for understanding through everyday experience, which may unconsciously increase tensions and frictions among these three main age classes.

This new trend is in clear contradiction to the very few old people at the top of the age pyramid during millions of years of hominisation. In fact, the recent change to a relatively narrow base and a much broader top on the age pyramid has happened in such a few generations that we cannot expect adaptations to this decisive change even in the fields of politics or sociology. The psychological consequences are still more difficult to predict and may be of much greater influence because selection can never adapt man on such an age distribution; and we cannot expect any innate modes of behavior for this new situation. This trend may become increasingly meaningful, especially for mass society.

Both of the areas discussed above emphasize clearly the problems arising from the far-reaching consequences of our characteristically human ability to change our environment and also from the increasingly accelerated move away from innate and learned modes of behavior, a move we presumably started long ago. Biologically speaking, we probably live in a very decisive and dangerous phase of exploring a new ecological niche, a niche that we are permanently and quickly changing by our behavior. This demonstrates that our recent modes of behavior are a very complex system of old innate modes and metabolic reactions combined with learned traits and self-stimulated changes. This is especially true today, given our modern age pyramid and population density.

SUMMARY AND CONCLUSIONS

The fossil remains of primates demonstrate clearly that some must have already changed from a predominantly arboreal to a partially or predominantly terrestrial life long before the Miocene. This process was probably favored by the fact that during the arboreal stage a remarkable reduction of the snout had already started, and, with this, the dominance of the nose for observation was replaced by the eye and hand to an increasing degree. The hominoids, the most interesting primates for our discussion, may have led a mainly terrestrial life in open areas for many millions of years before the surviving apes readapted more or less to an arboreal life.

For primates and particularly for hominoids, we presupposed a more or less large amount of functional bipedalism, known fully until now only by the hominids. During this process, the further freeing of the hands enabled at least the hominids to complete the transition from tool-using to toolmaking. Although we can state with confidence that primates already had a remarkable amount of intelligence, which is still increasing in the hominoids, the striking cerebralization of the hominids appears relatively late in the second half of the Pleistocene. This is confirmed not only by the capacity of the skulls but also by the inventory of transmitted cultural traits of the same period. From these facts, we must deduce that hominids passed nearly twenty million years of their evolution without specializations, so significant since about two million years ago and particularly during the last few hundred thousand years.

This leads to the conclusion that hominid survival was guaranteed in spite of the lack of special defense mechanisms such as large canines by a particularly close intragroup cooperation. This is already represented in other primates such as the Cercopithecoidea who offer in part better insights or models of our conceptions of the basal level of the hominids than the extant apes in their narrow ecological niches. For the intragroup cooperation and the therefore necessary social behavior, we already have in the Cercopithecoidea a combination of instinct and intelligence. The value of intelligence as a flexible basis for learning must have been particularly important for the hominids. The size of the groups in close cooperation probably never exceeded numbers which would exclude necessary personal contacts between all group members. For an open ecotype like Hominidae with worldwide distribution, the value of such small "social" units is particularly great. Only this can give an acceptable preliminary explanation for the fact that we find the outstanding habit of cannibalism as a "representative" characteristic in the known earliest human hominids. It emphasizes the loss of innate intraspecific barriers which could never be replaced by learned social controls as long as the hominids lived in small groups.

Additionally, life in such small units represented the only existing social environment for human hominids for nearly two million years. In the postglacial period, the beginning of the first intentional food production made it possible to surpass forever the limit of a few millions for the whole world population. But it is only in modern times with a population explosion caused by a technological civilization that we have millions of men in closest proximity. During this process, the small group, so long proved as effective for us biologically, lost its worth as a recognizable unit for the subconscious learning of social behavior. Given our fossil and subsequent history, it seems doubtful to expect the operation of innate modes for social behavior within personally unknowable masses. All observations until now seem to demonstrate that we never lived under special selection pressure in this direction. Also until now, the learned modes of behavior had the small group as a basis.

Biologically, the present population explosion, caused by our high degree of intelligence, seems one of the most difficult tests during the history of human hominids. Though we can change portions of our life conditions very rapidly and relatively decisively, the biological mechanisms for regulation seem to be directed mainly toward relatively slow, longtime changes. So now we have an outstanding gap between the biologically transmitted modes of behavior and the culturally learned modes. Since man acts far more emotionally than rationally, the question remains open how to get effectual measures for learning social behavior in a mass society without the opportunity of confirming those measures by adequate personal experience through direct contacts.

BIBLIOGRAPHY

BERGOUNIOUX, F. M., and F. CROUZEL
1965. Les Pliopithèques de France. *Ann. Paléont.*, 51:45–65.

DAVIS, P. R., and J. R. NAPIER
1963. A reconstruction of the skull of *Proconsul africanus*. *Fol. Primat.*, 1:20–28.

FEREMBACH, D.
1958. Les Limnopithèques du Kenya. *Ann. Paléont.*, 44:151–249.

HEBERER, G.
1959. Die subhumane Abstammungsgeschichte des Menschen. In G. Heberer (Ed.), *Die Evolution der Organismen*, 2. Aufl. Stuttgart: Fischer.

1965. Über den systematischen Ort und den psychophysichen Status der Australopithecinen. In G. Heberer (Ed.), *Menschliche Abstammungslehre*. Stuttgart: Fischer.

HÜRZELER, J.
1958. *Oreopithecus bambolii* Gervais. A preliminary report. *Verh. Naturf. Ges. Basel*, 69:1–48.

KELLNER, W.
1963. Ursachen eines hohen Krankenstandes. *Ärztl. Mitt.*, 60:2021–2030.

KOENIGSWALD, G. H. R. VON
1965. Die phylogenetische Stellung der Australopithecinen. *Anthr. Anz.*, 27:273–277.

KUMMER, B.
1965a. Das mechanische Problem der Aufrichtung auf die Hinterextremität im Hinblick auf die Bipedie des Menschen. In G. Heberer (Ed.), *Menschliche Abstammungslehre*. Stuttgart: Fischer.

1965b. Die Biomechanik der aufrechten Haltung. *Mitt. Naturforsch. Ges. Bern*, NF 22:239–259.

KUMMER, H., and F. KURT
1963. Social units of free living populations of Hamadryas baboons. *Fol. Primat.*, 1:4–19.

KURTH, G.
1965a. Die (Eu)Homininen. In G. Heberer (Ed.), *Menschliche Abstammungslehre*. Stuttgart: Fischer.

1965b. Die Bevölkerungsgeschichte des Menschen. In *Hdb. Biologie Akad. Verlagsanstalt Frankfurt*, IX:461–562.

216   GENETIC DIVERSITY AND HUMAN BEHAVIOR

Ludwig, W.
  1959. Die Selektionstheorie. In G. Heberer (Ed.), *Die Evolution der Organismen*, 2. Aufl. Stuttgart: Fischer.

Napier, J. R.
  1959. The problem of brachiation among the primates, with special reference to *Proconsul*. *Homo, Kongressband Kiel*, 187.

Napier, J. R., and P. R. Davis
  1959. The fore-limb skeleton and associated remains of *Proconsul africanus*. *Fossil Mammals of Africa*, No. 16. London: British Museum (Natural History).

Pilbeam, D. R., and E. L. Simons
  1965. Some problems of hominid classification. *Amer. Scient.*, 53:237–259.

Piveteau, J.
  1957. *Traité de Paléontologie*, T. VII. Paris: Mason et Cie.

  1961. Behavior and ways of life of the fossil primates. In S. L. Washburn (Ed.), *Social life of early man*. Chicago: Aldine.

Schultz, A. H.
  1960. Einige Beobachtungen und Masse am Skelett von *Oreopithecus* im Vergleich mit anderen katarrhinen Primaten. *Z. Morph. Anthr.*, 50:136–149.

Spatz, H.
  1962. Über Anatomie, Entwicklung und Pathologie des Basalen Neokortex. *Act. Med. Belg. Extr. Livr. Jub. du Dr. Ludo van Bongaert*, 766–779.

  1965. Vergangenheit und Zukunft des Menschenhirns. *Jb. 1964 Akad. Wissensch. u. d. Lit. Steiner/Wiesbaden*, 228–242.

Tobias, P. V.
  1965. Cranial capacity of the hominine from Olduvai Bed I. *Nature*, 208:205–206.

Washburn, S. L., and I. DeVore
  1961. Social behavior of baboons and early man. In S. L. Washburn (Ed.), *Social life of early man*. Chicago: Aldine.

Zapfe, H.
  1958. The skeleton of *Pliopithecus* (*Epipliopithecus*) *vindobonensis* Zapfe and Hürzeler. *Am. J. Phys. Anthrop.*, 16:441–457.

  1961. Ein Primatenfund aus der miozänen Molasse von Oberösterreich. *Z. Morph. Anthr.*, 51:247–267.

  1963. Fossile Menschenaffen im Wiener Becken. *Natur und Museum*, 93:295–404.

HUMAN GENETICS AND
THE THEME PATTERN OF HUMAN LIFE

FRIEDRICH KEITER

THIS PAPER presents a point of view that seems important for behavioral genetics but that certainly transcends the usual materials of morphology and behavior.

Inborn instinct in man has so often been claimed from the self-motivating themes of human behavior that we should try to make the pattern of human life themes themselves the object of a special study.

Very often "inborn instincts" in man have been stated without reference or assessment of respective neurological mechanisms. Such conclusions followed the reasoning that man must be motivated from "inside" wherever outward motivation was not obvious or well-understood. This argument was not without a certain merit in the beginning. But it was too rash in supposing that every "inside" motivation must be innate and genetic. It did not recognize the fact that behavior, whose motivation cannot easily be indicated, nevertheless can be the product of learning and conditioning. The "older instinctology" in anthropology therefore has been widely discredited.

But the most general ends and aim of human behavior are a fundamental anthropological problem in itself. They can be studied from the phenomena without any reference to neural or to unconscious background.

It is in cross-cultural comparisons that the malleability of human behaviors comes out best. So it seems of special importance to use the materials of cultural anthropology for working out the dimensions, the constancy, and the variation of the thematics of human life. By finding adequate metrics, perhaps we can give a definite measurable meaning and content to the commonplace that man has "unlimited" cultural modifiability. If we can get parameters for the average human behavior, we may also get a new base line for assessing the behavioral variation of man in terms of distances from these averages.

Thus our enterprise can be called a biological and quantifying parallel to the cultural-universals approach in culturology itself.

We try to develop the thematics of human life as deductions from the simple fact that man is a social primate with great ability for symboling and conceptual reasoning.

As we have seen, the argument of the older anthropological instinctology lacked stringency; behavior that is not obviously motivated from the outside is not thereby ascertained as instinctual and genetic. But there is not much doubt that a theme pattern of life processes which is deduced from man's being an animal, a primate, and a social, symboling, and conceptualizing being must have, directly or indirectly, a strong genetic background. We need not consider the question of genetic backgrounds at all if we only investigate what can be learned of the theme pattern of human life by systematizing the data of man's universal history, especially cultural history, along lines deducible from man's biological constitution.

Let us begin this systematization: obviously problems of the body, of the psyche, and of cultural-intellectual contents must co-work in every human life. So we get three *levels* of life themes: body-centered, psyche-centered, and culture-centered.

Following such distinctions, we must understand clearly that the whole man is *everywhere* in his behavior: he is an intellectual being when feeding, a sexual being when fighting, a social being when enjoying arts, and so forth. But the *stress* on different themes changes in human life from moment to moment, as the keys at the piano are always the same but always changed and combined in different ways by the player.

Within the three levels of body-centered, psyche-centered, and culture-centered behavior, *six thematic* distinctions are necessary on each level, and, if my many years' experience are correct, sufficient. So we get three times six = eighteen basic themes of human life.

These eighteen basic themes, deducible from the biological constitution of man, are found in every individual and collective human behavioral life; and there are no further phenomena in the thematics of human life that have remained unnoticed when taking these eighteen themes into account.

Most deductions for the biological necessity of the following basic themes of human life are so obvious that for the sake of brevity we will give an enumeration only:

Body-centered themes:
1. Metabolism
2. Natural environment
3. Life-cycle
4. Arousal and rest
5. Health and disease
6. Sex—eros and procreation

Psyche-centered themes:
7. Orientation
8. Inner states, emotions
9. Activities, challenges
10. Social environment
11. Positive and negative sociotropisms (group identification, helpfulness vs. freedom)
12. Dominance-submission

Culture-centered themes:

13. Language, knowledge
14. Cultural objects
15. Art
16. Norms and institutions
17. Ethos, existence philosophy
18. Transcendence, religion

Perhaps it is useful to give some examples of the biological deductions, which are behind the nomination of these, and only these, eighteen basic themes: (1) Life *is* metabolism wherever it exists. This fact cannot but have a bearing on the conscious behavior of man. (7, 8, 9) Psychical life *is* connection with the environment by means of sensory, neural, and muscular mechanisms of the organism; so orientation (cognition), inner states as reactions to life situations and activity in fighting challenges never can be absent from man's life. (18) Man's emotional reaction to experience transcends his objective cognition in range, so the help of dogma, cult, and magic never fails to be invoked. In this manner all our other themes also have very simple reasons and foundations.

I have set all my experience from working, travelling, and reading for many years to the task of finding out if these basic themes are sufficient and empirically adequate. A broad text in essay form especially devoted to this work is to be published as a book in Germany; I cannot go into details here.

In a strong scientific sense, the value of the system proposed here in brief depends on the possibility of developing *satisfying metrics* for the quantification of the facts and variables in question. This has been attempted with the following experimental materials.

First we attack the problem of the most extreme folkways set out by American cultural anthropology for the behavioral biology of man.

Alfred Kroeber long ago suggested a systematic collection of the extreme cases of folkways, mores, human customs, and behavioral regulations as a means for delimiting the potentialities of human nature. Kroeber was interested in the *limits* of human behavioral modificability. Certain other American cultural anthropologists, in contrast, suggested to the general reader a *nearly unlimited* and undefinable formative power of history and culture on man. Many of us got the impression that cultural learning and historical contingency could mold human behavior *ad libitum* indeed. A whole literature of books, of textbooks also, of social anthropology, of psychology, of sociology, and so forth never failed to bring impressive examples of this, and we have to be grateful for what we have learned from it. But as a corrective to this tendency the search for cultural universals began. Many deem the success of this approach as limited because it does not lead to sufficiently definite results. This is the criticism against which I try my quantifications.

For each of the eighteen basic themes parallel to Kroeber's suggestion (which was not known to me when I began this work), long lists of every extreme and, for westerners, nearly incredible folkway were compiled from all sorts of accessible sources. This compilation is not intended to be complete in an antiquarian sense, but is only a representative sample.

The first result of anthropological interest is that the contribution of each of the eighteen themes of human life to extreme folkways is not very different in amount. This could not have been foreseen. The broad variability of human behavior exhibited in history then is not a speciality of the culture-centered themes, but begins with regard to metabolism or with the elaboration of the life cycle.

What biological functions are behind extreme folkways? This of course is the question of most interest to us. We only can try a first and tentative survey of the factors, which can be assumed by such an analysis. For each item on our lists of folkways—that is, for 1,148 items for all eighteen themes—the following ten characterizations have been tried; the percentages give the frequency of the different supposed causal mechanisms:

I.	Unusual body functions	7.6%
II.	Unusual tolerance sensitivity	9.6
III.	Aberrant cognitive situations	21.2
IV.	Values that are odd to westerners	27.5
V.	Extraordinary achievements of will power	4.5
VI.	Constraint by fellow men	2.1
VII.	Direct consequences of economic poverty and opulence	2.2
VIII.	Consequences of the geographic environment	1.6
IX.	Evolutionary escalation, reaching extremes step by step	5.7
X.	Organismic constitution of man rather than life situations causally prominent	18.0

As one can see, geography, economic opulence, and constraint by fellow men only rarely were obvious. Evolutionary escalation has an interesting frequency; much odd behavior is the consequence of departing from the average step by step. Often enough we can learn from cultural anthropology that the human body is capable of improbable functioning and that tolerance for pain and stress can exceed by far our western ideas and standards.

But the most important source of extreme folkways is to be seen in strange cognitive situations and "odd" valuations. Here man's intellectual capacities and the openness of man's values to change come into play.

All in all, extreme folkways do not teach us as much as one could expect about biological functions in man's behavior; much of what we learn already has been known to western psychologists and has been treated in traditional psychological textbooks. The excitement caused by cultural anthropology nevertheless has its good reason. If anthropological studies of the peoples and cultures of the world had not been done, we could not have guessed from western experience alone what abundance of behavioral "architectures," regulations, and stabilizations is implicit in these known mechanisms.

Category X lists all extreme folkways that seem to have direct bearing on the organismic constitution of man in general. These 18 per cent are 207 single items, every one of which is amenable and liable to special psycho-dynamic investigation in the field and in the lab.

For a tentative elaboration of the statistic parameters of the thematics of human life, a *content analysis with thirty samples of the universal narrative literature* of the world has been tried.

Human life, as a total process, is understood by men themselves, and is codified in the narrative literature of all peoples and ages. A uniform treatment of such materials, all in German or English translation, seemed to me a rewarding approach to a quantitative assessment of the thematic content of human life. The inherent difficulties and limitations of this assessment have become known full well during this work, but all in all the result is sufficiently good.

Every one of the eighteen basic themes was split into a convenient number of secondary themes; thus natural environment, for instance, was divided into climate, landscape, plants, catastrophes, light, and so forth. A dual code indicating basic and secondary themes was made. Then on the margin of the printed texts these code signs were noted to express the essential topics or themes of the ongoing tale. A sample of 1,000 code signs were collected from each source for statistical treatment.

Some of the literature included was from Australians, Melanesians, Polynesians, and other ethnological groups; from old Egypt, old China, Plutarch, Augustinius, *Fanny Hill*, different medieval texts; and from anthologies of short stories from many peoples of the present world. The experiment was designed to get information on different ages, different parts of the world, and different types of culture and style simultaneously.

Every one of the eighteen basic themes is represented in all thirty samples. Nowhere was it necessary to force data into narrowly prescribed channels; these basic themes proved appropriate. So the primary requirements for a working system of the thematics of human life for our biological purposes seem fulfilled. The validity of our count can be seen further from the fact that all frequencies of the themes in the narrative literatures occur in normal distributions.

The historical modificability of the narrated behaviors is not unlimited. All thirty frequencies always remain within the range of $+ 3$ sigmata, distance from the mean. The following statistical parameters indicating the *average human behavior as reflected in the narrative literature* may be of interest (see table on page 222). All figures are pro-mille!

The frequencies and variances of the themes are quite different as we are wont to see in anthropometric data, and such like. Higher coefficients of variation, V, indicate more change in history. Thus philosophy and religion are labile, but references to the life cycle or to the social environment are stable or more invariant in history.

What has been counted here from literary texts could be abstracted by an adequate metric from direct field observation also. The Harvard New Guinea Expedition of 1961 is an example of field work which has been reported not by

	M	σ	V
1. Metabolism	32	11	34
2. Natural environment	64	25	39
3. Life cycle	35	9	26
4. Rest	27	8	30
5. Disease	18	8	29
6. Eros, procreation (sex)	44	19	43
7. Orientation	75	24	32
8. Inner states, emotions	90	31	34
9. Activities and challenges	120	30	25
10. Social environment	84	23	27
11. Sociotropisms	56	20	36
12. Dominance-submission	69	29	42
13. Language, knowledge	48	20	41
14. Cultural objects	96	25	26
15. Art	28	12	43
16. Institutions, norms	50	22	44
17. Ethos, existence	25	16	64
18. Transcendence, religion	47	32	68

abstract statements, but by a thorough-going story of events by a scientist of high literary ability. We could analyze this extraordinary scientific report like narrative literature!

Secondary themes are much less frequent, of course, than the basic themes whose specifications they are. For secondary themes the statistical frequency distribution is therefore often skewed. Many samples show low frequencies while some few luxuriate in especially high frequencies. It is an old ethnological experience that cultural activities are often concentrated in a few groups and are more or less lacking in all others. Here we have an interesting statistical expression of this fact that deserves further study.

We can characterize the behavior of every human individual or every human group by assessing the intensity of the basic and secondary themes *in relation to the pan-human averages* given above in a first tentative approximation. Personal and national characterology must profit from such overall reference to average man.

Ranking of our thirty literary samples on a time dimension provides a contribution of our work to the problems of behavioral evolution in man.

We distinguish three strata: ethnology, history, and modern literature. It can be called consistent evolution if the frequency of one of our themes is increasing or decreasing from the first to the third of these strata. There are of course other possibilities, for example, invariance over all three strata or climax in the middle (historic) stratum. About half of all themes checked (including secondary themes) show consistent evolution in the above sense. This is a broad enough foundation for stressing the soundness of evolutionary thinking in our field.

In another way our data have given interesting information. We have summed up the frequencies of all single themes relating to the same general phenomenon. So all single frequencies that indicate aggression, politico-military interest, sex, pain, pleasure, emotion, esthetical interest, intelligent behavior, higher education, and so forth have been pooled for every one of our thirty samples. By this procedure we find an extraordinarily stable ratio of 66:33 per cent for indications of pleasure to pain. It is of anthropological interest that modern opinion polls found this same ratio for self-evaluations. References to aggression and bloodshed are becoming rarer and rarer; the reference to intelligence has not increased from the historic (middle) stratum onward, reference to intellectual education, to the state, to policy, to heroism also has its climax *before* modern times. Philosophy is very unstable in frequency; religion dwindles but never is lacking totally in any literature.

Findings on national character on the level of thematic analysis are difficult to summarize briefly.

The most general result here is the high positive correlation between the sum of thematic differences of two literatures and their distance in space and time.

This result is also, as may be mentioned, a check on the validity of our procedure. It is not unexpected perhaps, although Kroeber concluded any continuous drifting away of cultural forms in space and time would be counterbalanced by diffusion (see his famous tree of life vs. tree of culture). Metrics for the total difference of two cultural systems are not so easily found that our experiment could not be of sure interest.

We mention a few more special results. Modern short stories of African and Asian authors are extremely different in thematic content and show we do not have a unified world civilization that has lost all differentiation. Rather the newly invented medium of the art of writing short stories gives the differences in old traditions new expression.

The comparison of old Australia and the modern United States shows unexpectedly little reference by the narrators to differences in technology in the two cultures; indeed, they show much less difference in life-thematics than would be expected in such extremely different civilizations.

All in all, modern and ethnological narration shows human life in its essentials to be much more similar than could be expected. Our method evidently is effective in stripping off the changing shell of humanity and laying bare its lasting invariant core.

Growing differentiation in the thematics of human life is evidently not on the level of basic themes, but on the level of secondary themes anyway. The percentage of code signs that are *not* used in a sample is lowest in the modern short story and highest in the Australian material.

The reference to technical objects does not show much historical change; only in the semi-nomadic Australians is it definitely less frequent than in all other groups.

Our most similar groups are Melanesians, Polynesians, and Micronesians. Samples from the same people in different epochs show considerable historic insta-bility, though never a decided dissimilarity. In Germany the comparison of dif-ferent literary fashions of our own century is of interest.

Four different samples on bellicose aristocracies show similar pattern and dif-ferent backgrounds side by side (Schach-Name, San Kwo Tchi, Plutarch, Ice-land).

As a triangle with extreme differentiation of the pattern of life themes, the heroes of Plutarch, the saint Augustinus, and the English prostitute Fanny Hill have been compared. The sexual theme even in *Fanny Hill* appears in only 18 per cent of the items. Only within limits can the human life process be distorted by extreme valuation of any single one of its themes.

The human life process is a pattern of many interrelated variables. Our basic themes and still more the secondary themes are, from the point of view of corre-lation and factoral analysis, mostly uncorrelated. They are the factors that might have been discovered by statistical analysis, whereas I simply followed experience. The "average man" and his regular variations are the biological reality of *Homo sapiens* itself. Evidently, there is not a superposition of more than one thematical pattern, existing within humanity. It can be shown that small and big deviations from the general averages are distributed over our thirty samples in a random manner, notwithstanding some consistencies stemming from evolution or history.

If I am correct, this approach has not floundered in its intention to give a first sketchy quantification of the unity and variation of human behavior.

The search for other metrical clues must be continued. In behavioral genetics it can be merit enough to find out exactly that here and there genetical back-grounds do *not* exist. Perhaps this merit is to be given to all sound statements on psychological race differences. Thematic analysis also can find genetical back-grounds only where they exist.

Two things are clear. First, there is not any specification of the human life process that could *not* also be expressed in terms of thematic analysis. Children, women, men, neurotics, geniuses, asthenics and athletes, gifted and dull people—they can all be characterized by surplus or deficit of the themes of their life against all-human averages. Most national characters must have characteristics on this metric also.

Second, the general theme pattern of the life of *Homo sapiens* depends broadly, *evidently and without doubt*, on genetical constitution. Is it not totally unlikely, then, that genetic influence and genetic background should stop short with the general species characteristics of *Homo sapiens* without working into the indi-vidual, typical and collective populational variations of humanity also?

SUMMARY

A system of eighteen basic and ten times as many secondary themes is deduced with logical necessity from the fact that man is a social, symboling, conceptualizing primate. With this system the extreme folkways as indicators of the limits of man's historical modificability have been studied. Perhaps new psychomechanisms were not found, but an extremely broad field of possible regulations and stabilizations is produced by the known mechanisms. Quantitative paramaters have been obtained by a thematic content analysis of thirty samples from the world's narrative literature. The basic themes of human life are completely ubiquitous; they vary within fixed limits in the form of normal distributions. The thematic pattern of human life shows historical evolution. National characters or other collective qualities of groups have been studied with this new metric. Humanity can be shown to follow one general pattern in the themes of the life processes, the variations oscillating randomly around the pan-human averages. Many genetically determined traits of human behavior must be supposed to be ascertainable with these parameters, but it is dubious if any such facts have already been found.

BEHAVIOR AND MORPHOLOGICAL VARIATION[1]

GARDNER LINDZEY

THE THEME of this sermon is that human structure and function are significantly intertwined. I propose to assert the relative neglect by American students of behavior of the physical person and its components, to suggest some possible determinants of this persistent myopia, to examine some arguments and evidence for the importance of such constitutional parameters for the psychologist, and, finally, to mention a few illustrative areas of potential research interest. Initially I should make clear that I am using the term morphology quite broadly to refer not only to the physical, structural aspects of the organism but also to any externally observable and objectively measurable attribute of the person, thus embracing such variables as hirsuteness, symmetry, color and even esthetic attractiveness.

AMERICAN RESISTANCE TO CONSTITUTIONAL PSYCHOLOGY

It is a commonplace observation that American psychologists have been reluctant to give serious consideration to the study of morphology and behavior (see MacKinnon and Maslow, 1951; Hall and Lindzey, 1957), and some of the determinants of this enduring resistance are quite clear. First, it is undeniable that we have devoted our attention primarily to the study of social and behavioral change. The modal emphasis among psychologists in America has been upon learning, acquisition, shaping, or the modification of behavior, and not upon those aspects of the person and behavior that appear relatively fixed and unchanging. As has been said elsewhere,

One important by-product of American democracy, the Protestant ethic, and the dogma of the self-made man has been the rejection of formulations implying that behavior may be innately conditioned, immutable, a "given." Because it is commonly accepted that physical characteristics are linked closely to genetic factors, the suggestion that physical and psychological characteristics are intimately related seems to imply a championing of genetic determinism. It is not surprising that such a conception has been unable to muster much support in the face of the buoyant environmentalism of American psychology (Hall and Lindzey, 1957, p. 337).

1. A modified version of this paper was presented as the presidential address of Division 8 (Personality and Social Psychology) of the American Psychological Association, September 5, 1964. The manuscript was written while in residence at the Center for Advanced Study in the Behavioral Sciences and its preparation was facilitated by a grant from the Ford Foundation, Grant MH-11030 from the National Institute of Mental Health, and a Special Fellowship from the National Institute of Mental Health. I am grateful to Edward E. Jones for his helpful comments.

A second significant deterrent has been the spectacular failure of the influential formulations of Gall and Spurzheim (1809) concerning body and behavior, coupled with the popularity of morphological variables among such diagnostic charlatans as palmists and physiognomists. These two conditions provide a ready-made clock of naivete, ignorance, or dishonesty for the person interested in working in this area.

Third is the fact that constitutional psychology has in recent decades become so closely associated with the name and work of William Sheldon that attitudes toward the one are scarcely separable from attitudes toward the other. In his research and writing Sheldon is much more the sensitive naturalist, observer, and categorizer and much less the hard, quantitative, and objective scientist than would be optimal to assure a good press from our colleagues. Moreover, in his writings he has proven to be singularly adept at ridiculing or parodying just those aspects of the scientific posture of psychologists that are most sensitively, rigidly, and humorlessly maintained. One might argue convincingly that, if Sheldon had conducted the same research but had reported it in an appropriately dull, constricted, and affectless manner (consistent, let us say, with *Journal of Experimental Psychology* standards), its impact upon the discipline of psychology might have been much greater. What I am suggesting is that acceptance of Sheldon's work was impaired by a general resistance on the part of American psychologists to the study of behavior in relation to fixed characteristics and, conversely, that his irreverent and unconventional style provided further support for the belief that the study of morphology and behavior was an unsanitary practice.

Whatever the determinants, it seems clear that American psychology in the past and present has maintained a suspicious hostility toward formulation and investigation concerned with body and behavior. Perhaps only the study of behavior in relation to the soul has been equally unpopular among psychologists.

Morphological Dimensions

A word should be said concerning available variables and methods for describing the physical person. It is obvious that we do not yet have agreement concerning those components or dimensions that can be employed most fruitfully to represent morphological variation, although there are three sets of variables and attendant measures that have had a reasonable range of modern application. First, and best known, are Sheldon's (1940, 1942) components of endomorphy, mesomorphy, and ectomorphy, with ratings for each dimension derived from a standardized set of photographs by means of a complex rating procedure or through the actuarial use of a small number of relatively objective indices. The full details of the latter procedure have not yet been fully published, but the indices are age, height, ponderal index, and trunk index. Parnell (1958) has devised an alternative method of measuring comparable variables which he labels fat, muscularity, and linearity. His ratings are based upon a small number of anthropometric measures of subcutaneous fat, bone length, and girth of arm and calf. Third is Lindegard's (1953; Lindegard and Nyman, 1956) scheme which includes

four variables—length, sturdiness, muscularity, and fat. His dimensions were identified with the aid of factor analysis, and they are rated by means of a combination
of anthropometric measurements and performance (strength) tests.

It is clear that much important research remains to be done in further identifying alternative or additional dimensions to represent the body and that the existing
dimensions and measures are far from ideal. Indeed there have been a number of
publications (for example, Hammond, 1957a, 1957b; Hunt, 1952) that have been
sharply critical of the best known measurement systems. Still, the deficiencies
that inhere in these particular schemes and measurement operations must be expected to attenuate or obscure attempts to relate these variables to behavior. Thus,
whatever well-controlled associations we now observe between morphology and
behavior may be considered to represent minimum estimates rather than an upper
limit. It is important to note that, in spite of diversity in the particular measures
or indices investigators have commenced with and differences in the methods of
analysis they have employed, there is a high degree of congruence among the
three sets of variables that are currently popular. Finally, and perhaps most important, we do possess a set of existing tools with which one readily can initiate research in this area.

Mechanisms Mediating a Hypothetical Relationship between Morphology and Behavior

Ultimately, justification for increased interest in morphology and behavior
must rest upon the provision of compelling empirical evidence for important
associations between these two sectors of the person. Before turning to even a
glimpse at empirical evidence, however, it seems reasonable to examine the rational and theoretical basis that can be provided to motivate an interest in such
relationships. In brief, what we are concerned with here is a specification of the
reasonable pathways or mechanisms whereby an important association between
physical and behavioral components might be achieved. A crude classification
suggests five discriminable means that might lead one to predict such associations.

First is the possibility that a common experiential or environmental class of
events has a characteristic influence upon personality or behavior and, at the same
time, has a regular and detectable effect upon morphology. For example, if we
accept certain psychoanalytic formulations (Bruch, 1947) concerning the relation between obesity and a particular type of parent-child relation—maternal overprotectiveness—and assume, moreover, that physical components are influenced by
weight and diet factors, we would confidently expect some degree of association
between the endomorphic or fat factor and those behavioral consequences that
are believed to be associated with maternal overprotectiveness. Or, if we accept
the report of Landauer and Whiting (1964) that in societies where noxious and
traumatic experiences are a regular part of the socialization practice, "the mean
adult male stature was over two inches greater than in societies where these customs were not practiced" and assume, moreover, that such early experiences also
have an influence upon behavior, we would once again expect physique and be-

havior to be importantly linked. There is, of course, a good deal of lower animal evidence (for example, Denenberg, 1962; Lindzey, Winston, and Manosevitz, 1963) to support this same point of view—that early experience influences both morphology and behavior. In brief, this position makes the reasonable assumption (with some supporting evidence) that certain important events serve to determine both physical and behavioral development in a manner that produces significant associations between outcomes in these two spheres.

Second is the inevitable observation that behavior is directly limited or facilitated to some degree by the physical person. Even the most dedicated and competitive 145-pound athlete cannot aspire realistically to play first-string tackle for a Big Ten university, nor is it likely that the 260-pound tackle could ever compete successfully in a marathon race or as an effective jockey. A male whose maximum height is 5′4″ cannot reasonably expect to compete with John Thomas or Valeri Brumel. The frail ectomorph cannot expect to employ physical or aggressive responses with the same effect as the robust mesomorph. Height, strength, weight, and comparable dimensions place direct and unmistakable limits upon what responses the individual can hope to make adaptively in a given environmental setting.

A third factor, which is closely related to the second, concerns the indirect consequences of a particular set of physical attributes. For example, within our society, if an individual's biological makeup places him well above most others in size and strength, it is very likely that he will be recruited early in his life into competitive athletics and that this experience will play an important role in his life. Those who have undergone or studied the impact of four to ten years of varsity athletics upon behavior would scarcely question the profound implications of this experience for many important aspects of the individual's values, dispositions, and overt behavior.

To take a widely divergent example, it is now carefully documented for women that linearity (ectomorphy) is negatively associated with rate of physical and biological maturation. A study by McNeil and Livson (1963), the only published research with which I am familiar that employs Sheldon's new and presumably more actuarial method of somatotyping, provides compelling evidence for a significant association between a process of great psychological importance and one morphological component. No observer of the adolescent female or student of the socialization process would be likely to deny that the age at which a girl becomes, in some sense, a woman has significance for her psychological development; and here we have clear evidence that the more linear her physique, the slower she will be in undergoing and completing this developmental stage. Consistently, a series of investigations conducted as part of the Oakland Growth Study (Eichorn, 1963; Jones and Bayley, 1950; Jones, 1965) has demonstrated clear and consistent behavioral differences between individuals who vary in their rate of physical maturation.

The several examples that have been cited could be multiplied endlessly by listing the implications of particular kinds of bodies for particular kinds of en-

vironmental events and experiences. Surely there is some degree of regular behavioral impact attendant upon being a handsome man as opposed to an ugly
man; a "stacked" girl as opposed to a fragile, linear girl; a large, powerful man
in comparison to a small, weak man.

Fourth is a special case of the indirect implications of physique, and relates
to role specification or the social stimulus-value of particular variations in the
physical person. Insofar as a given society includes a number of individuals who
hold common expectations in regard to the "fat man," the "lean and hungry
person," "the red head," "the receding chin," and so forth, it is clear that the
individuals who fit these physical prescriptions will be exposed to a somewhat
different set of learning experiences or environmental events than the person of
more modal body or whose physique is extreme in a different manner. Whether
the behavior that is expected of the particular physique is grounded in firm empirical observation, casual superstition, or magical thinking, if the expectation is
maintained by a sizable number of persons it creates a different social reality
for those who fit the prescription. Whether they conform to social expectation
or vigorously oppose it, they are influenced by a common set of experiential determinants that set them off from those of a different morphology; consequently,
we might expect to find some uniformities in their behavior not shared by persons
of very different physique.

Finally, we come to the mechanism which is probably most objectionable to
the majority of American psychologists—joint biological determinants of both
behavior and physique. This type of association could be produced by a known
set of physiological processes that demonstrably influence both behavior and
physique (in the abstract it is easy to cite endocrine function), or the link might
be produced by a common set of genetic determinants, with or without information concerning the process whereby genetic variation is translated into morphological and behavioral variation. To take a simple case, for both man and lower
animal it is well known that a large proportion of the morphological variables
that have been studied are importantly influenced by genetic variation. For
example, the two major twin studies that have examined physical attributes
(Newman, Freeman, and Holzinger, 1937; Osborne and DeGeorge, 1959) report
heritability coefficients for specific morphological attributes that for the most
part are above .70. Sheldon's physique components are reported by Osborne and
DeGeorge to display appreciably lower heritability coefficients (Total Somatotype: Males $= .36$; Females $= .61$), but it remains clear that genetic variation
makes a significant contribution to morphological variation.

Given this observation, it seems altogether reasonable to anticipate that the
chromosomal loci that influence variation in morphology may have multiple or
pleiotropic effects, some of which are behavioral. To use an illustration made
somewhat remote by phylogenetic regression, my colleague Harvey Winston and
I (Winston and Lindzey, 1964), working with mice, have produced evidence
suggesting that albinism (which is determined by a recessive gene cc in linkage
group 1) is associated with a relative deficiency in escape learning. Whether this

particular finding is sustained or not, there is every reason to expect that comparable pleiotropic effects linking physical and behavioral components will be found in a variety of behavioral and physical areas. Even at the human level we may expect comparable, although perhaps more complex, examples. Indeed, if we are willing to include the realm of pathology in our discussion, it is clear that there are numerous examples of conditions that, under the normal range of environmental variation, are controlled by genetic variation and include both behavioral and physical deviations. An evident example of this is Down's syndrome (mongolism) which, as a result of modern cytological advances, we now know to be produced by a genetic anomoly, typically involving either trisomy or translocation associated with the 21st chromosome (Jarvik, Falek, and Pierson, 1964), and which leads to a variety of dramatic behavioral and physical consequences. Here, then, is a particular kind of genetic variation (objectively identifiable) that leads to profound and undeniable effects in both the physical and behavioral sphere. Numerous other examples could be discussed, including Huntington's chorea, phenylketonuria, and infantile amaurotic idiocy.

The main generalization I wish to extract from what has just been said is that, working within the assumptions that are common to most psychologists (whether learning, developmental, physiological, social, or what not), it appears altogether reasonable for one to expect to observe important associations between morphology and behavior. Or, stated in the ritual terminology of modern psychology, given appropriate information concerning morphology we may expect with reasonable confidence to have some degree of predictive control over the variance of behavior.

Some Illustrative Findings

This is obviously not the place to attempt a serious literature survey, but I can select several areas where there has been a reasonable amount of investigative activity and examine the trend and implications of the data. The two active areas I will discuss are concerned with (1) estimating the direct relation between observer ratings of personality and morphological variables, and (2) relating morphology to a complex achievement variable, in this case criminal behavior.

Personality Ratings. Almost every introductory psychology student has learned to recognize and distrust the striking findings that Sheldon (1942) reported between dimensions of physique and temperament. Studying 200 male college students individually over a considerable period of time, he assigned ratings to each subject for his three somatotype components and also, on the basis of extensive interviewing and observing, made ratings for three temperament variables that were intended to represent aspects of behavior associated with each of the dimensions of physique. When the temperament ratings were correlated with ratings for the associated physique component, the correlations ranged from $+ .79$ to $+ .83$. A number of factors would have to be considered in a full discussion of why so much covariation was observed, but for most psychologists the explanation has seemed to lie in the fact that Sheldon himself executed both sets of ratings. Consequently, one may reason that implicitly Sheldon's prior convictions or expecta-tions in this area led him to rate both physique and temperament in a consistent manner, whatever may have existed in reality.

Many other studies have related morphological variation to ratings of person-ality, and most of them have produced at least some evidence for association between these two domains. Let us here direct our attention to two studies that have minimized, more successfully than most, the possibility of contamination of data or experimenter bias.

The first investigation was executed by Irvin Child (1950), utilizing 414 male, undergraduate students who were routinely somatotyped as freshmen and who, as sophomores, completed a questionnaire that consisted of multiple-choice ques-tions concerning the subjects' own behavior and feelings. Each item was designed to measure some aspect of behavior that, on the basis of Sheldon's study of tem-perament, was believed to be related to one or more dimensions of physique. In advance of any analysis of the data, a total of 94 predictions were made concern-ing associations between self ratings and somatotype ratings—these predictions were simple reflections of what Sheldon had already reported. Sheldon assigned the somatotype ratings on the basis of the standard photographs and without any knowledge of the questionnaire results. More than three quarters (77 per cent) of the predictions were confirmed in regard to the direction of the relationship between physical component and self rating, and more than one-fifth of the pre-dictions were confirmed at the 5 per cent level, while only one of the findings that reversed the prediction was significant at the 5 per cent level. When Child combined certain of the items into scales for each of the temperament variables so that they could be correlated with the physique dimensions, he observed posi-tive correlations between the comparable physique and temperament dimensions, the highest of which was + .38.

The most important observation here is that even when we eliminate the major design and inference shortcomings of Sheldon's study, we still observe a pattern of relations between physique and behavior that resembles in direction, if not de-gree, the findings reported by Sheldon. The magnitude of association between self ratings and morphology is obviously much less than that reported by Sheldon, but it is difficult to know what proportion of this change is attributable to elim-ination of experimenter bias and what proportion is due to the relative insensi-tivity of self ratings.

A similar study was carried out by Walker (1962), working with 125 male and female nursery school children. The somatotype measures were based upon ratings made independently by three different judges, which were then combined by an averaging method. Two of these judges never saw the children while the third was acquainted with them. Each of the children was also rated on a 1-to-7 scale for 64 specific behavioral items by two to five judges, none of whom were in-volved in the morphological ratings and at least two of whom were ignorant of the purpose of the study. Again, the individual score for each item was deter-mined by averaging the judges' ratings. These items also had been selected for their pertinence to Sheldon's findings, and a total of 292 specific predictions were made in advance and tested separately for the male and female subjects. Altogether, 73 per cent of the predictions were confirmed in direction and 21 per cent were confirmed at or below the 5 per cent level of significance, while only 3 per cent were disconfirmed at the 5 per cent level.

Once again we observe significant and appreciable relations between morphological variation and behavior that are consistent with Sheldon's findings, even though the major opportunities for systematic bias have been removed. Again, the degree of association is less than that reported by Sheldon, but the fact that these observer ratings of behavior might be considered less sensitive than those used by Sheldon, in addition to the use of young children as subjects, makes it difficult to consider these findings as clearly contradictory on this score.

The principal generalization I wish to derive from these findings is that the most firmly based evidence we now possess suggests the existence of important associations between morphology and behavior (just as reason would assert), while the magnitude of this relation, or rather its varying magnitude depending upon the conditions of study, remains to be determined precisely. If, as the climate of current opinion urges, the magnitude of association reported by Sheldon represents in large measure covariance attributable to experimental error, this fact remains to be demonstrated unequivocally. Moreover, in view of the extensive criticism of his study, it does seem odd that there has not been one single effort at a careful replication eliminating the major defects in Sheldon's study, while at the same time attempting to preserve other relevant conditions as exactly as possible. What we have witnessed, instead, has been the complacent dismissal of a potentially important set of results with no serious attempt at an empirical resolution.

Criminal Behavior. We turn now to a morphological correlate that is far removed from specific ratings of items or components of personality. Here we are concerned with a complex outcome variable, criminal behavior, that includes a wide variety of topographically different forms of behavior and is undoubtedly related to many different antecedent events. Again I will limit my discussion to the initial findings reported by Sheldon and a small number of subsequent investigations involving similar questions and methods.

Over an eight-year period Sheldon and his collaborators (Sheldon, 1949) conducted a study of physique and behavior with the residents of a rehabilitation home for boys in Boston. The findings of this study are largely clinical or descriptive but do include a graphic comparison of college youths and delinquent subjects that makes clear that while the college distribution shows a clustering about the mid-range physique (4-4-4), the delinquent youths show a tendency to cluster in the "northwest region." In brief, mesomorphs, particularly endomorphic mesomorphs, are over-represented among the delinquents. This generalization, although initially supported by little in the way of convincing empirical data, has been reexamined subsequently in four separate studies that have involved better controls and more objective analysis of data.

The best known of these subsequent studies was conducted by Glueck and Glueck (1950, 1956) and involved the study of 500 persistent delinquents and 500 proven nondelinquents, matched in age, intelligence, ethnic background, and place of residence. Independently of any knowledge of the classification of the individual subjects, their physiques were rated in terms of the relative dominance of Sheldon's three components. An examination of the somatotype distributions for the two groups provides a powerful confirmation of Sheldon's findings: approximately 60 per cent of the delinquent youths were classified as mesomorphic while only about 30 per cent of the nondelinquent subjects were so classified. Moreover, less than 15 per cent of the delinquents were categorized as ectomorphic and almost 40 per cent of the normal subjects were placed in this category. A final item of confirmation was found in the incidence of endomorphic mesomorphs in the delinquent (13 per cent) and nondelinquent (3 per cent) subjects. All in all, under what appear to be very well controlled circumstances, a quantitative analysis of the relation between criminal behavior and morphological variation provides a highly significant and impressive confirmation of Sheldon's rather impressionistic findings.

A study of Epps and Parnell (1952) is of particular interest because it involved both a change in society (England) and in sex (girls) and yet also led to the observation that when female delinquents were compared with female college students they were "shorter and heavier in build, more muscular and fat" (p. 254). A more recent study of Gibbens (1963) also employed English subjects, in this case male, and again found a substantial confirmation of the finding that delinquent youth are predominantly of "northwest physique."

Another recent study (Cortes, 1961) involved the use of Parnell's techniques of measurement and the comparison of 100 adolescent boys who had been convicted by courts of violations of the criminal law with a group matched in age but with no record of delinquent behavior. When the subjects were classified according to their dominant physique component, Cortes found 57 per cent of the delinquents to be mesomorphic while only 19 per cent of the normal subjects were so classified. On the other hand, 33 per cent of the normal subjects were dominantly ectomorphic in comparison to 16 per cent of the delinquents. In spite of the different somatotyping method, and other variations in procedure, these results almost exactly parallel the findings of the Gluecks and provide further confirmation for the original association reported by Sheldon.

Thus, four separate studies involving different methods, different sexes and different societies, some of them employing excellent controls against experimenter bias, have produced findings that are consistent among themselves and congruent with Sheldon's initial report. No one is likely to argue seriously that mesomorphy directly causes (or is caused by) criminal behavior, but an association as consistently and strongly observed, and involving as complex and socially important a variable as criminal behavior, obviously warrants further systematic study.

It is even possible to see hints of an association between these findings and those that have been reported in connection with psychopathic or sociopathic traits. In an ingenious study, Lykken (1957) reported a deficiency in capacity for avoidance learning on the part of the "constitutional psychopath." More recently, Schachter and Latané (1964) replicated this finding and introduced the fascinating additional observation that sociopaths show considerable improvements in avoidance learning when injected with adrenaline, while normal subjects demonstrate no such change in performance. It is interesting to note that Sheldon's (1942) early description of the somatotonic temperament (corresponding to mesomorphy) placed considerable emphasis upon indifference to pain and low anxiety. If it should eventually be shown unequivocally that low anxiety, deficient avoidance learning, and an atypical effect of adrenaline upon avoidance learning are all associated with extremes of mesomorphy, it would be possible to integrate a variety of discrete and individually important results bearing upon criminal behavior. Such results obviously would have implications extending far beyond study of the psychopath or criminal.

The implication of genetic factors in this relationship between criminality and physical components is made reasonable by the findings of twin studies concerned with criminal behavior. An early study by Lange (1928) compared monozygotic and dizygotic twins in their concordance in having been imprisoned. In the case of monozygotic twins, ten of thirteen were concordant (the second twin had also been imprisoned), while in the case of dizygotic twins only two of seventeen pairs showed a similar concordance. Lange accepted this difference between fraternal and identical twins in concordance rate as an indication of the important role of heredity as a determinant of criminal behavior. Eysenck (1964), who also argues strongly for genetic component in criminal behavior, surveyed the existing literature and found a total of 225 pairs of twins who had been studied by various investigators with respect to adult crime. The average concordance rate for the identical twins was 71 per cent, while the fraternal twins showed a concordance of only 34 per cent, suggesting appreciable heritability for this attribute.

Areas of Potential Research Interest

I have just mentioned several areas of investigation that seem to me to promise significant returns. I should like to allude briefly to a few additional topics that may offer similar rewards. It should be clear that these selections are quite arbitrary and in no sense are intended to represent the entire range of investigative possibilities.

Effects of Early Experience. A recent consequence of Freud's seminal observations and the findings of ethologists and comparative psychologists has been an intense interest in the effects of early experience. Numerous studies with both human and lower animal subjects exploring this problem (Bowlby, 1951; Denenberg, 1962) have convinced most observers that variation in early experience has important implications for adult behavior or personality. Unfortunately there are many disagreements concerning the exact nature of the function linking infantile experience with adult behavior. Even in connection with such a seemingly simple question as whether traumatic or strong noxious stimulation in infancy has adaptive or maladaptive effects, or increases or decreases emotionality, the answer seems by no means clear (for example, Denenberg, 1961; Levine, 1961; Hall and Whiteman, 1951; Lindzey, Lykken, and Winston, 1960, 1961a, 1961b).

One significant question here is whether there may not be important differences in the nature (and perhaps even the direction) of effects from the same noxious stimulus when administered to different classes of organisms; whether the same stimulus may not have a quite different impact upon different bodies, with different sensitivities and varying response capacities. It is at least possible that certain kinds of stimulation will have divergent effects upon the fragile, linear, hyper-reactive infant and the heavy, spherical and sluggish infant. Indeed, one might argue that evidence from a variety of lower animal sources (King and Eleftheriou, 1959; Valenstein, Riss, and Young, 1955), including our own research (Lindzey, Lykken, and Winston, 1960; Lindzey, Winston, and Manosevitz, 1963; Winston, 1963), indicates that the effects of early experience are influenced by genotype or biological variation, thus providing presumptive support for just such a formulation. All of this implies that introduction of morphological variation as a parameter or variable to be studied when examining the effects of early experience might shed considerable light upon the complexities of this important area.

Psychopharmacology. To select still another fashionable area of research, there is the largely unexamined interaction between morphological variation and effects of psychoactive drugs. One of the major puzzles that plagues research in psychopharmacology is the observation of very large individual differences in reaction to the same doses of the same drug under the same environmental conditions. Undoubtedly many factors contribute to this variation but one such class may consist of morphological variables. Indeed, beginning with Sheldon's (1942) temperament descriptions of his polar physical varieties we find a number of hints that observers have believed this to be the case. In fact, the gross morphological variable of weight has always been considered an important parameter in predicting the effects of any drug—psychoactive or not. Moreover, there is considerable evidence from lower animals for the influence of biological determinants, including genotype, upon response to particular drugs, including alcohol (McClearn and Rodgers, 1959, 1961). All in all, whether concerned with alcohol, narcotic agents, or the more recent hallucinogenic and tranquilizing drugs, there seems every reason to expect morphological variation to play a meaningful role in determining the psychological effects of the drug.

Social Interaction and Morphology. As a final illustrative item it may be appropriate to mention a topic that appears to have little fit with current psychological activities. Modern social psychology has shown very little interest in the physical structure of the persons who are studied in temporary or enduring interaction. Indeed, an examination of the index of the *Handbook of Social Psychology* (Lindzey, 1954) reveals no entries for "physique" or "somatotype" and only a single entry for "physical factors."

In spite of this systematic neglect, few, if any, would deny that the physical attractiveness of a person, whether male or female, has a marked influence upon the responses he evokes from others. It would come as a shock to most of us if we found that social status, popularity, and such had no relation to esthetic appeal. Likewise the questions of how big, how muscular, or how well formed an individual may be have obvious implications for the way in which he is perceived by others, as well as the way in which he participates in group functioning.

Many years ago Freud (1914) pointed to a fascinating area of investigation in his remarks concerning the social significance of the narcissism encountered prototypically in the big cats and beautiful women. For a complex of reasons we seem to have shown little professional interest in lions, tigers, or stunning women. Perhaps now is the time to restore beauty and other morphological variables to the study of social phenomena.

BIBLIOGRAPHY

Bowlby, J.
1951. *Maternal care and mental health.* Geneva: World Health Organization.

Bruch, Hilde
1947. Psychological aspects of obesity. *Psychiatry,* 10:373–381.

Child, I.
1950. The relation of somatotype to self-ratings on Sheldon's temperamental traits. *J. Pers.,* 18:440–453.

Cortes, J. B.
1961. *Physique, need for achievement, and delinquency.* Unpublished doctoral dissertation, Harvard University.

Denenberg, V. H.
1961. Comment on "infantile trauma, genetic factors, and adult temperament." *Psychol. Rep.,* 8:459–462.
1962. The effects of early experience. In E. S. E. Hafez (Ed.), *The behaviour of domestic animals.* Baltimore: Williams and Wilkins.

Eichorn, Dorothy H.
1963. Biological correlates of behavior. In H. W. Stevenson (Ed.), *Yearb. Nat. Soc. Stud. Educ.,* 62 (1):4–61.

Eysenck, H. J.
1964. *Crime and personality.* Boston: Houghton-Mifflin.

Epps, P., and R. W. Parnell
1952. Physique and temperament of women delinquents compared with women undergraduates. *Brit. J. med. Psychol.,* 25:249–255.

Freud, S.
1914. On narcissism. In *Standard Edition Vol. XIV.* London: Hogarth Press, 1957.

Gall, F. J., and J. G. Spurzheim
1809. *Recherches sur le Systeme nerveux.* Paris: Schoell.

Gibbens, T. C. N.
1963. *Psychiatric studies of Borstal lads.* London: Oxford University Press.

Glueck, S., and Eleanor Glueck
1950. *Unraveling juvenile delinquency.* New York: Harper.

1956. *Physique and delinquency*. New York: Harper.

HALL, C. S., and G. LINDZEY.

1957. *Theories of personality*. New York: John Wiley and Sons.

HALL, C. S., and P. H. WHITEMAN

1951. The effects of infantile stimulation upon later emotional stability in the mouse. *J. comp. physiol. Psychol.*, 44:61–66.

HAMMOND, W. H.

1957a. The constancy of physical types as determined by factorial analysis. *Human Biol.*, 29:40–61.

1957b. The status of physical types. *Human Biol.*, 29:223–241.

HUNT, E. A.

1952. Human constitution: an appraisal. *Amer. J. phys. Anthrop.*, 10:55–73.

JARVIK, LISSY F., A. FALEK, and W. P. PIERSON

1964. Down's syndrome (Mongolism): the heritable aspects. *Psychol. Bull.*, 61:388–398.

JONES, MARY C.

1965. Psychological correlates of somatic development. *Child Develpm.*, 36:899–912.

JONES, MARY C., and NANCY BAYLEY

1950. Physical maturing among boys as related to behavior. *J. educ. Psychol.*, 41:129–148.

KING, J. A., and B. E. ELEFTHERIOU

1959. Effects of early handling upon adult behavior in two subspecies of deermice, *Peromyscus maniculatus. J. comp. physiol. Psychol.*, 52:82–88.

LANDAUER, T. K., and J. W. M. WHITING

1964. Infantile stimulation and adult stature of human males. *Amer. Anthrop.*, 66:1007–1028.

LANGE, J.

1931. *Crime as destiny*. London: Allen and Unwin. (Originally published in 1928.)

LEVINE, S.

1961. Discomforting thoughts on "infantile trauma, genetic factors, and adult temperament." *J. abnorm. soc. Psychol.*, 63:219–220.

LINDEGARD, B.

1953. Variations in human body-build. *Acta Psychiatrica et Neurologica*, Supplementum 86:1–163.

LINDEGARD, B., and G. E. NYMAN

1956. Interrelations between psychologic, somatologic, and endocrine dimensions. *Lunds Universitets Arsskrift*, 52(8):1–54.

LINDZEY, G. (Ed.).

1954. *Handbook of social psychology*, Vols. I and II. Cambridge, Mass.: Addison-Wesley.

LINDZEY, G., D. T. LYKKEN, and H. D. WINSTON

1960. Infantile trauma, genetic factors, and adult temperament. *J. abnorm. soc. Psychol.*, 61:7–14.

1961a. Confusion, conviction, and control groups. *J. abnorm. soc. Psychol.*, 63:221–222.

1961b. Trauma, emotionality, and scientific sin. *Psychol. Rep.*, 9:199–206.

LINDZEY, G., H. D. WINSTON, and M. MANOSEVITZ

1963. Early experience, genotype and temperament in *Mus musculus. J. comp. physiol. Psychol.*, 56:622–629.

Lykken, D. T.

1957. A study of anxiety in the sociopathic personality. *J. abnorm. soc. Psychol.*, 55:6–10.

McClearn, G. E., and D. A. Rodgers

1959. Differences in alcohol preference among inbred strains of mice. *Quart. J. Stud. Alcohol.*, 20:691–695.

1961. Genetic factors in alcohol preference of laboratory mice. *J. comp. physiol. Psychol.*, 54:116–119.

MacKinnon, D. W., and A. H. Maslow

1951. Personality. In H. Helson (Ed.), *Theoretical foundations of psychology*. New York: Van Nostrand.

McNeil, D., and N. Livson

1963. Maturation rate and body build in women. *Child Develpm.*, 34:25–32.

Newman, N. H., F. N. Freeman, and K. J. Holzinger

1937. *Twins: a study of heredity and environment.* Chicago: University of Chicago Press.

Osborne, R. H., and F. V. DeGeorge

1959. *Genetic basis of morphological variation.* Cambridge, Mass.: Harvard University Press.

Parnell, R. W.

1958. *Behavior and physique: an introduction to practical and applied somatometry.* London: Arnold.

Schachter, S., and B. Latané

1964. Crime, cognition and the automatic nervous system. *Neb. Symp. Motivation.* Lincoln: University of Nebraska Press.

Sheldon, W. H.

1940. (with the collaboration of S. S. Stevens and W. B. Tucker) *The varieties of human physique: an introduction to constitutional psychology.* New York: Harper.

1942. (with the collaboration of S. S. Stevens) *The varieties of temperament: a psychology of constitutional differences.* New York: Harper.

1949. (with the collaboration of E. M. Hartl and E. McDermott) *Varieties of delinquent youth: an introduction to constitutional psychiatry.* New York: Harper.

1954. (with the collaboration of C. W. Dupertius and E. McDermott) *Atlas of men: a guide for somatotyping the male at all ages.* New York: Harper.

Valenstein, E. S., W. Riss, and W. C. Young

1955. Experimental and genetic factors in the organization of sexual behavior in male guinea pigs. *J. comp. physiol. Psychol.*, 48:397–403.

Walker, R. N.

1962. Body build and behavior in young children: I. Body build and nursery school teachers' ratings. *Monogr. Soc. Res. Child Develpm.*, 27(3) (Serial No. 84).

Winston, H. D.

1963. Influence of genotype and infantile trauma on adult learning in the mouse. *J. comp. physiol. Psychol.*, 56:630–636.

Winston, H. D., and G. Lindzey

1964. Albinism and water escape performance in the mouse. *Science*, 144:189–191.

BEHAVIOR AND MATING PATTERNS
IN HUMAN POPULATIONS

J. N. SPUHLER

THE PATTERN OF MATING is one way that determines how genes are combined in genotypes in human and other bisexual populations. Mutation, gene flow, selection, and genetic drift usually are considered to be a complete list of the systematic modes of change of gene frequencies in such populations. Given the gene frequencies for a generation and knowledge of the pattern of mating, we can in principle predict the distribution of genotypes for that generation.

Thus different patterns of mating may have different behavioral consequences in a breeding population to the extent that differential behavior is in some part a result of differences in genotypes. The causes of differential patterns of mating in human populations are not understood in any detail, although it is certain that the causes involve a complex interplay of biological, demographic, social, and cultural factors (Ford, 1945; Lorimer, 1954).

The paths between biology and behavior may run in both directions and circle back; we know that cultural differences operating on the pattern of mating may result in changes in the distribution of genotypes in a breeding population and that these genetic changes may lead to differences in the acquisition and performance of at least some items in the cultural repertory by individuals.

We will consider three general patterns of mating: random mating, inbreeding, and assortative mating.

In no human population is the mating perfectly random. If a breeding population has N males, and if a given female has a first child by a given father, the probability under strictly random mating that her next child will have the same father is $1/N$. The observed frequency of this event is closer to unity than to $1/N$ in all known human populations. However, in many cases where the theory of population genetics is used to interpret observed genotype frequencies, the departures from randomness in mating are small enough to be negligible. This is true, for instance, of the distribution of blood-group genotypes in several hundred known populations (Mourant, 1954).

The idea of random mating as used in the Hardy-Weinberg equilibrium (where the array of genotypes in a randomly mated population is given by the square of the gene frequencies when these are expressed as a binomial or multinomial summing to unity) provides a convenient reference point for analysis of changes in the distribution of genotypes resulting from inbreeding or assortative mating.

INBREEDING

Inbreeding is the mating of individuals who have one or more biological ancestors in common.

The two most important observed consequences of inbreeding in experimental and farm animals are (1) the reduction of the mean phenotypic value shown by characters connected with reproductive capacity or physiological efficiency, and (2) the increase in uniformity, or reduction in the variance, about the mean phenotypic values within inbred lines. The reduction in fitness, termed *inbreeding depression*, may decrease the mean in fertility of *Drosophila* (per pair per day) as much as 1.25 per cent for each 1 per cent increase in inbreeding; in mice, litter size at birth may be decreased by 0.8 per cent, and female weight at six weeks of age by 0.3 per cent per 1 per cent increase in inbreeding (Falconer, 1960). As will be shown in the material to be reviewed below, inbreeding depression is well established for morphological, physiological, and behavioral characters in human populations.

Both inbreeding depression and uniformity within inbred lines find a fully verified genetic explanation in the increased homozygosis consequent upon inbreeding. Thus, as Wright (1921) first suggested, it is desirable that an inbreeding coefficient should run on a scale from 0 to 1 while the percentage of homozygosity is running from 50 to 100 per cent (50 per cent being the proportion of homozygosis for two alleles with equal frequency in a randomly mated population), and the coefficient should measure as directly as possible the consequences of inbreeding to be expected on the average from any given regular or irregular pedigree.

In the sections to follow, the properties of Wright's coefficient of inbreeding (which he named "F" because intense inbreeding results in the "fixation" of genes in homozygous genotypes, that is, it brings about the complete absence of heterozygosis) will be explored in sufficient detail to interpret some observed results on inbreeding in human populations. Attention also will be called to inbreeding effects on genotypic variance and on the genetic correlation of biological relatives.

The methods used by anthropologists and biologists to measure the degree of inbreeding before Wright (see Pearl, 1915) were based on the intuitively reasonable idea of "ancestor loss," the fact that an inbred individual possesses fewer different ancestors in some particular generation than the maximum possible number, which is 2^n for the nth ancestral generation. This method is defective and sometimes quite misleading because the same degree of "ancestral loss" is found for certain patterns of mating known to have radically different inbreeding consequences and to result in different degrees of homozygosis.

The operation of a pattern of mating gives a lineage of breeding populations a structure in time consisting of a complicated network of parent-child relationships. The net (Fig. 1) consists of individual organisms (father, mother, son, daughter, sperm, egg, zygote) connected by paths of two sorts. A single-headed arrow indicates the direct effect along that path of one variable upon another. For example, the genotype of a zygote is completely determined by the sperm and the egg that unite to form it. A double-headed arrow indicates residual relations going back to factors not included in the diagram. For example, in Figure 1, $\overset{M}{\longleftrightarrow}$ stands for the correlation between mates, and $\underset{F}{\longleftrightarrow}$ for the correlation between uniting gametes. These two correlations are of great importance in the study of mating patterns, for M is used to measure the degree of assortative mating and F to measure the degree of inbreeding. F and M will be considered in greater detail below. Their full evaluation requires the tracing of paths more remote than those shown in Figure 1.

Methods of Path Coefficients

The method of path coefficients, an invention of Sewall Wright, is perhaps the most convenient tool for the genetical analysis of such networks; Wright (1921, 1951) used this method to develop a major portion of the genetic theory of systems of mating. We will consider here only those parts of the theory needed to introduce Wright's conclusions on the genetic consequences of inbreeding and assortative mating.

If $\overline{V}_0$ is the mean and σ_0 is the standard deviation of the variable, V_u, the standardized form, X_0, of that variable is $X_0 = (V_0 - \overline{V}_0)/\sigma_0$. Assume that X_0 is a linear function of known variables $X_1, X_2, \ldots X_m$, and that the causal network is made formally complete by addition of statistically independent hypothetical or unanalyzed variable, x_n. Single-headed arrows in the network may be indi-

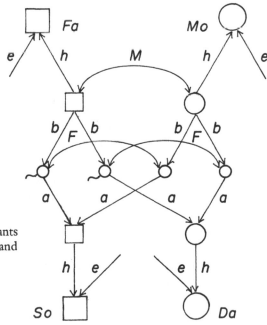

FIGURE 1
Path diagram showing determinants of parent-parent, parent-child, and sib-sib correlations (after Wright, 1921).

cated by path coefficients, P_{o1}, representing the path from X_1 to X_o. The relations between X_o and the known and hypothetical variables may then be written $X_o = P_{o1}X_1 + P_{o2}X_2 + \ldots P_{om}X_m + P_{on}X_n$.

The correlation, r_{oq}, between two standardized variables, X_o and X_q, is defined as their mean product $r_{oq} = \sum_{i=1}^{k} P_{oi}r_{iq}$. The correlation of a variable with itself, $r_{oo} = 1$, leads to the equality $\sum_{i=1}^{n} P_{oi}r_{oi} = 1$, which may be partitioned into a component $\sum_{i=1}^{m} P_{oi}r_{oi}$ expressing the portion of the variance (σ^2_o) due to known factors—in other words, the squared coefficient of multiple correlation—and a component $P_{on}r_{on} = r^2_{on}$, the portion of the variance due to unknown factors.

The paths for genes passing along the network of biological descent are represented in Figure 1. Variation in the phenotypes of individuals are assumed to be determined by the additive effects of a hereditary (h) and an environmental path (e) so that $e^2 + h^2 = 1$. Variation in the genotypes of individuals is determined completely by the gametes (over paths a) that unite to produce them. The path coefficient a relates zygotic genotype to one of these gametes. The genotypic value is assumed to be the sum of the values of the two gametes that produce it. For each of these gametes we add the two paths aa and aFa so that $2a(a + aF) = 1$ and $a =$

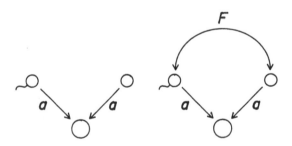

$\sqrt{[(1/2)/(1+F)]}$. Variation in gametes formed by individuals is determined over path b, which relates a gamete to the genotype that produced it. This path coefficient b is also the correlation between that gamete and the genotype, since the two are connected by a single path. With regard to the alleles at a specific locus, that correlation expressed by b is equal to the correlation between that genotype and one of the gametes a generation back that produced that particular genotype. Using primes to indicate the preceding generation, $b = a' + a'F' = \sqrt{[(1+F)/2]}$.

With these definitions, we may now find the correlation between parent and parent, r_{pp}, between parent and offspring, r_{po}, and between brother and sister, r_{oo}; these correlations are of great importance to behavior genetics because they can be estimated directly from observations on family members. A single correlation, M, connects parent and parent. There are two paths connecting parent and offspring: $habh + habMh = h^2ab(1 + M)$. Four paths of two kinds connect brother and sister: $2(habbah) + 2(habMbah) = 2h^2a^2b^2(1 + M)$. It should be noted that M

$(= F/b^2)$ is taken to add up all possible connections between the parental geno-types. Under panmixia $M = F = O$ and r_{po} and r_{oo} reduce to 0.50 b^2.

The b^2 in the above path equations measures the "heritability" of a trait with reference to a specific population and environment. In this context it is the ratio of the additive genetic variance to the total phenotypical variance: the term is used with a variety of other statistical meanings, especially in twin studies.

Much biological variation of interest to behavior genetics is likely to involve nonadditive connections between genotype and phenotype due to gene interaction within (called dominance) and between (called epistasis) loci. And in human be-havioral traits there are nearly always some nonadditive relations between geno-type and environment. The complexities due to nonadditivity are difficult to handle and are discussed in works on quantitative genetics (Wright, 1951; Kemp-thorne, 1957; Falconer, 1960).

CALCULATION OF F FROM PEDIGREES

The value of F can be obtained by tracing the parent-child paths that connect an inbred individual to all ancestors common to his two parents. The method is illustrated in Figure 2 showing the matings of (a) brother and sister, and (b) single first cousins.

Starting with Figure 2(a), consider the transmission of the four genes A^1, A^2, A^3, A^4, at some autosomal locus in the two common ancestors, that is, the parents

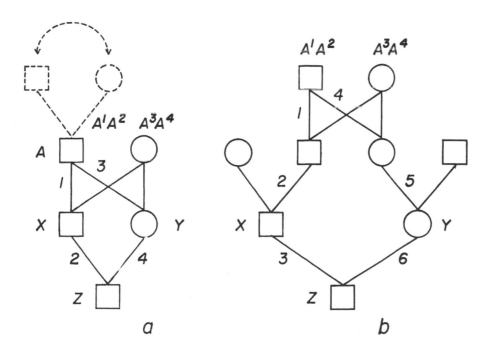

FIGURE 2

Pedigrees of an inbred child from (a) a brother X sister mating, and (b) a mating of single first cousins. One of the common ancestors in pedigree (a) is inbred.

of X and Y. The probability that gene A^1 will reach the inbred child, Z, over parent-child steps 1 and 2 is $(1/2)^2 = 1/4$; the probability that the same gene will reach Z over steps 3 and 4 is also 1/4. Steps 1 and 2 together with steps 3 and 4 form one of two inbreeding loops connecting Z with his two common ancestors. The probability that A^1 will reach Z over both sides of one inbreeding loop is 1/16, which is the probability that Z will be homozygous A^1A^1 for two genes identical by descent. The same argument holds for genes A^2, A^3, and A^4. Thus the probability that Z will be homozygous for genes identical by descent, that is, for any one of the four genes in question, is by definition the inbreeding coefficient of offspring of brother $\times$ sister matings; this probability is $4 \times 1/16 = 1/4$.

In the case of the child whose parents are single first cousins, Figure 2(b), the probability that gene A^1 will reach the inbred child Z over parent-child steps 1, 2, and 3 is 1/8. Thus the probability that Z will be homozygous for A^1A^1 is $(1/8)^2$ and that of homozygosity for any one of the four genes present at an autosomal locus in the two common ancestors is $4(1/8)^2 = 1/16$.

If a common ancestor is inbred, as indicated by the dashed pedigree in Figure 2(a), the two genes A^1 and A^2 in that common ancestor may be identical by descent from a more remote ancestor common to his parents; the probability of this is, by definition, the inbreeding coefficient of the more proximate common ancestor, say, F_A. The fact that A is inbred increases the probability that X and Y will receive genes identical by descent from A by a factor $F_A/2$, and all inbreeding loops connecting to an inbred common ancestor require multiplication by the factor $(1 + F_A)$.

By extension of these procedures, the contribution of any common ancestor, A, to the inbreeding coefficient of a descendant individual, Z, is $(1/2^n)(1 + F_A)$ where n is the number of individuals in the loop leading from one parent to the common ancestor and back through the other parent. An inbred individual may have more than one common ancestor, and more than one inbreeding loop may connect an inbred individual and the same common ancestor. The contribution of all common ancestors to the inbreeding coefficient of a descendant is the sum of the contribution of all inbreeding loops connecting all common ancestors:

$$F = \Sigma[(1/2)^n(1 + F_A)]$$

where n is the number of individuals in the inbreeding loop not counting Z.

If the genes being considered are sex linked, only female offspring may be inbred (since, barring non-disjunction, males get a Y chromosome from their fathers and are hemizygous for sex-linked genes), only females in the inbreeding loop are counted—omitting the inbred females whose inbreeding coefficient is desired—and any loop containing a father-son step is omitted entirely.

The above formula allows us to find the inbreeding coefficients of individuals, sibships, members of a generation, or members of a breeding population. It may be extended to pedigrees of any degree of complexity, as suggested in Figure 3, the pedigree of the most inbred sibship in the Ramah Navaho Indian population (Spuhler and Kluckhohn, 1953).

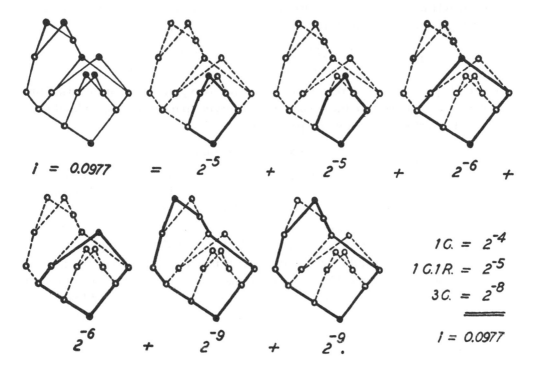

$$i = 0.0977 \quad = \quad 2^{-5} \quad + \quad 2^{-5} \quad + \quad 2^{-6} \quad +$$

$$2^{-6} \quad + \quad 2^{-9} \quad + \quad 2^{-9}.$$

$$1\,C. = 2^{-4}$$
$$1\,C.1\,R. = 2^{-5}$$
$$3\,C. = 2^{-8}$$
$$\overline{}$$
$$i = 0.0977$$

FIGURE 3

Calculation of the inbreeding coefficient of a child whose parents are related as first cousins, first cousins once removed, and third cousins.
(From Spuhler and Kluckhohn, 1953.)

For comparatively highly inbred, natural, and enduring human populations, application of the formula given above for F is tedious because of the difficulty of finding and counting all of the inbreeding loops once and only once. Wright and McPhee (1925) found a sampling method for approximating the inbreeding coefficient of a population. Kudo (1962) and Kudo and Sakaguchi (1963) suggested an exact method for calculating inbreeding coefficients which is easily adapted to machine programming.

F AS THE CORRELATION OF UNITING GAMETES

The above results suggest an increase in the proportion of homozygous among autosomal loci is, on the average, an expected population consequence of inbreeding, and we can use these results to explore a second definition of the inbreeding coefficient, that is, F = the correlation between uniting gametes. The array of genotypes for two autosomal alleles in a breeding population may be written in terms of the gene frequencies (where p is the frequency of A and $(1-p)$ that of the other allele, a) and the proportion of heterozygotes, H:

$$[(p - H/2)\text{AA} + H\text{Aa} + (1 - p - H/2)\text{aa}].$$

The correlation, F, between the gametes that united to form this array of genotypes is given by the formula for a four-fold table

$$F = \frac{(p - H/2)[(1-p) - H/2] - (H/2)^2}{\sqrt{[p^2(1-p)^2]}} = \frac{H_o - H}{H_o}$$

where $H_o \ [= 2p(1-p)]$ is the proportion of heterozygotes under panmixia. Thus F may be used to measure the deviation of genotype frequencies in an inbred population from that expected under panmixia.

Panmixia and Fixation

Now, if we define a coefficient of panmixia, P, as $P = (1 - F)$, we can summarize in compact form the results to this point in Table 1 which gives the frequencies of genotypes from two autosomal alleles under three patterns of mating: (1) random mating or panmixia, where $F = O$ and $P = 1$; (2) an intermediate degree of inbreeding, where $O < F, P < 1$; and (3) complete fixation, where $F = 1$, $P = O$, and heterozygotes are completely absent from the population. The intermediate condition of mating may be expressed in three equivalent ways: (1) deviation from panmixia; (2) a compound of panmictic and fixed components; and (3) deviation from fixation. It is important to note these changes in genotype frequency take place without change in gene frequency; they may be reversed by changes in the pattern of mating.

TABLE 1

Genotype frequencies under three patterns of mating
for autosomal alleles with frequencies.

Mating Pattern		AA	Aa	aa	Totals
Panmixia $F = 0$		p^2	$2pq$	q^2	1
Intermediate $O < F < 1$	Deviation from panmixia	$p^2 + Fpq$	$2pq - 2Fpq$	$q^2 + Fpq$	$1 + 0$
	Panmictic and fixed components	$Pp^2 + Fp$	$2Ppq$	$Pq^2 + Fq$	$P + F = 1$
	Deviation from fixation	$p - Ppq$	$2Ppq$	$q - Ppq$	$1 - 0$
Complete fixation $F = 1$		p	$-$	q	1

Recurrence Relations for F

We have considered the F values for isolated cases of brother $\times$ sister and single first-cousin mating. Recurrence equations can be obtained in cases where regular patterns of mating occur over several or many generations. These equations have been obtained for regular patterns of close inbreeding by working out in algebra the consequences of every possible type of mating and giving each type its proper weight in each generation. For example, Jennings (1916) showed the decrease in heterozygosis (H) for a regular system of brother $\times$ sister mating is given by the terms of the Fibonacci series

$$\frac{1}{1}, \frac{1}{2}, \frac{2}{4}, \frac{3}{8}, \frac{5}{16}, \frac{8}{32}, \frac{13}{64}, \ldots$$

where the numerator is the sum of its preceding two terms and the denominator doubles every term, giving the recurrence equation $H_{n+2} = (1/2)H_{n+1} + (1/4)H_n$.

Any method of working out all possible mating types is not practical for more complex systems of mating; for example, the case of double first-cousin matings with autosomal linkage involves the consideration of 10,000 different pairs of mating types. However, Wright's method of path coefficients leads quickly to results identical with the tedious, if more elegant, algebraic methods (Wright, 1921, 1923, 1951).

The regular practice of continued sib mating breaks a population into separate and branching lines of descent consisting, at any given generation, of "isolates" whose members are a sibship. Since only brothers and sisters are mates, the correlation (M)—see Figure 1—between mating individuals in generation n is that between full sibs in generation n-1; that is, $r_{pp} = r'_{oo}$, or

$$m = 2a'^2(b'^2 + b'^2m')$$
$$= 2a'^2(b'^2 + F').$$

Substituting the values of a'^2 and b'^2, we find

$$m = \frac{1 + 2F + F''}{2(1 + F')}$$

or

$$F = b^2m = (1/4)(1 + 2F' + F'').$$

Since $H = H_0(1\text{-}F)$, the recurrence equation for H is

$$H = H_0(1 - \frac{1}{4} - \frac{1}{2}F' - \frac{1}{4}F'')$$
$$= \frac{1}{2}H_0(1 - F') + \frac{1}{4}(1 - F'')$$
$$= \frac{1}{2}H' + \frac{1}{4}H''$$

which is identical with the recurrence relation $H_{n+2} = \frac{1}{2}H_{n+1} + \frac{1}{4}H_n$ obtained from the Fibonacci series.

If we assume the decay of heterozygosis process has reached equilibrium by $H/H' = H'/H''$, then $H/H' = (1/4)(1 + \sqrt{5}) = 0.809$, indicating heterozygosis decreases about 19.1 per cent per generation under continued brother $\times$ sister mating. Wright has shown the heterozygosis decrease rate in a population of size N with random mating (union of gametes) is $1/2N$ per generation.

By using the method of path coefficients, the effects of continued regular systems of inbreeding may be obtained for single first and second cousins and other mating patterns of anthropological interest as shown in Figure 4.

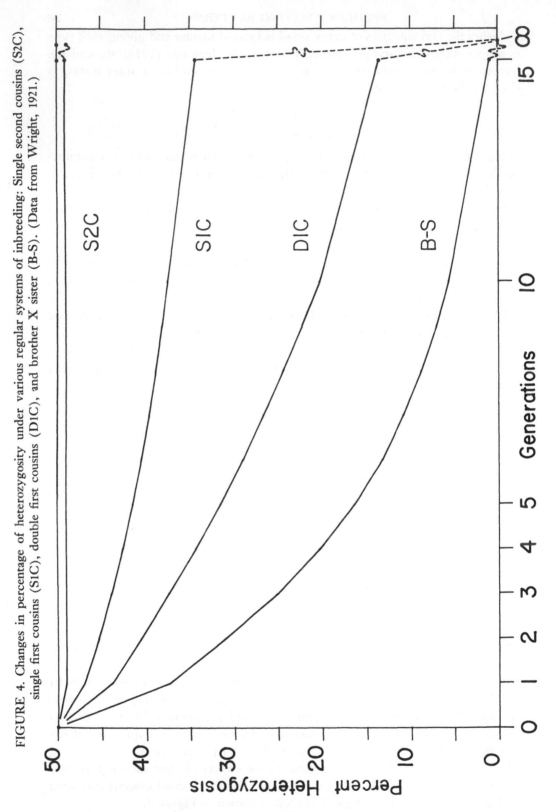

FIGURE 4. Changes in percentage of heterozygosity under various regular systems of inbreeding: Single second cousins (S2C), single first cousins (S1C), double first cousins (D1C), and brother X sister (B-S). (Data from Wright, 1921.)

250

The genetic consequences of inbreeding implied by some particular F value must be taken as relative to some reference population; the statement that the coefficient of inbreeding is X per cent in population A at time T_1 implies that heterozygosis is X per cent less than expected from random mating within the founding or base population of, say, time T_o to which the pedigrees trace. If we let the subscript I refer to individuals; G to groups, strains, isolates, or other subdivisions of the population; and T to the total, base, or founding population, then F_{IT} is the inbreeding of individuals with reference to the total population; F_{IG} that of an individual relative to a group within the comprehensive population; and F_{GT} that relative to the base population that would persist if random mating within the subpopulations were initiated. The mating structure of populations may be analyzed by application of the formula $(1 - F_{IT}) = (1 - F_{IG})(1 - F_{GT})$, where F_{GT} is necessarily positive but F_{IG} and F_{IT} can be negative.

Degree of Inbreeding in Human Populations

The mean F values for known human populations occupy only a small part of the theoretical range from 0 to 1. The Samaritans, with $F = 0.043$, have the highest degree of inbreeding reported for a human group (Bonné, 1963). The highest F value reported for a human individual, 0.4270, was calculated by Rasmuson (1961) for Cleopatra-Berenike III—an aunt of the Cleopatra we associate with Caesar and Antonius—on the basis of the unreliable pedigree drawn up by Strak in 1897, which probably mistakes unrelated women given the term of address and reference "sister" after marriage for biological sisters.

The mean F for the general population of France in the period 1926–1945 was 0.00066 (Sutter and Tabah, 1948) and that of England and Wales in the period 1924–1929 was 0.00028. Bunak (1965) suggests many contemporary rural populations of Europe, North America, and Negro Africa are composed of demes numbering 1,500 to 4,000 individuals, with about 80 per cent endogamy, and with a preponderant part of the deme members related to one another as third cousins, that is, with $F = 0.0039$. A number of human geneticists take the maximum value commonly observed in isolated populations, about 0.006, as the probable long-term F value for the human species (see, for example, Neel *et al.*, 1949). Morton, Chung, and Mi (1967) used data from a study of interracial crosses in Hawaii to estimate the inbreeding coefficient due to remote consanguinity in large populations is about 0.0005. Wright (1950) arrived at a value of 0.02 for the long-term inbreeding coefficient in human populations by considering random mating for a neutral gene within local breeding populations of 200 mated couples. The available ethnographic data suggest this is an underestimate of the upper limit to the inbreeding coefficient in primitive human populations, for it is improbable that human bands exceeded a total of 200 to 300 persons of all ages (or 50 to 75 breeding pairs) during the history of the species before agriculture. Known food-gathering people without domestic animals other than the dog rarely live in groups over fifty or sixty individuals (Linton, 1955). If this is indeed the general case, various enduring local populations of our species passed through an "inbreeding bottleneck," after a very long time in transit, from a few to not more than some 300 generations ago.

INBREEDING AND BEHAVIOR

Hundreds of investigations on the consequences of inbreeding for behavior, especially for the defective and gifted extremes of the behavioral spectrum, appeared prior to the development of population genetics in the first third of this century. Huth (1875) compiled the earlier studies, Westermarck (1903) brought together the data from the last quarter of the nineteenth century, and East and Jones (1919) surveyed the information from the opening decades of the tweniteth century. Holmes (1924) gave an extensive bibliography of these inquiries, nearly all of them now lacking other than historical interest.

Despite extended statistical scrutiny, the behavioral outcome to be expected on the average from the marriage of near kin remained controversial—some experts were for and others against it. The following explanation for this lack of agreement is quoted here from East and Jones (1919) because it is still widely accepted in anthropological circles:

The impossibility of a correct statistical answer to the problem is clear if one works back from the answer given by research on heredity: "Inbreeding is not in itself harmful; whatever effect it may have is due wholly to the inheritance received." It is not to be wondered, therefore, that examination of the pedigree record of one family led to one conclusion, and of another family to exactly the opposite (pp. 243–244).

East and Jones, *Inbreeding and Outbreeding*, appeared two years before Sewell Wright, *Systems of Mating*, 1921, the starting point of modern population genetical studies on inbreeding.

INBREEDING AND BEHAVIOR IN JAPANESE CHILDREN

A study of Japanese children is up to now the most elaborate attempt to measure the behavioral consequences of inbreeding in a human population. The study was organized by the Human Genetics Group at the University of Michigan under the direction of J. V. Neel in collaboration with a number of Japanese workers and carried out by the Child Health Survey, Hiroshima and Nagasaki, in 1958–1960 (Schull and Neel, 1962, 1965). I had the pleasure of being field director for the survey in 1959 but did not take part in the analysis of the data by W. J. Schull and Neel. I am indebted to these colleagues for making available an advanced copy of their analytical results.

I will give a brief account of the Hiroshima results. Schull and Neel (1965) may be consulted for full details on the study in both cities. The sample numbers of inbred and control boys and girls available for neuromuscular and psychometric observation in Hiroshima were:

	Males	*Females*	*Totals*
Inbred (F > 0)	552	959	1,511
Controls (F = 0)	621	987	1,608
Totals (F ≥ 0)	1,173	1,946	3,119

The behavioral variables observed in Hiroshima under the direction of Arthur L. Drew were:

1. and 2. "Age when walked" and "Age when talked" are two behavioral criteria of common use in pediatric and parental appraisals of child development. They are, of course, subject to errors of parental recall and may be biased by cultural norms.

3. A dynamometer was used to measure strength of grip in kilograms in the right and left hands. The handle of the instrument was adjusted to fit the hand size of each child who was allowed several practice tries.

4. Tapping rate is the average number of taps with the index finger on a telegraphic-type key over five periods, each of ten seconds duration. The rate is recorded separately for the two hands.

5. The Color Trail is a paper-and-pencil test with five pairs of simple geometric figures (triangles, circles, and so forth), each pair of two different colors, the same color being present in one member of two different figures in the case of three pairs and one color being unique to one of the other two pairs. The object of the test (which is easier to do than to describe) is to connect a figure of one color to the same shape of another color, then to a dissimilar shape of the same color, then to the same shape of another color, and so on. Each child had a practice run on a simplified version. A maximum of 120 seconds was allowed. The score is the number of seconds taken for completion.

6. The maze test has five pencil mazes of the Porteus type that are included in the Japanese version of the Wechsler Intelligence Scale for Children (WISC). Three maze trails were available for practice. Each maze must be completed in a specified time; if more mistakes than a specified number are made, the score is zero points; if the maze is completed with exactly the allowable number of mistakes, the score is 1 point. Scores of 2 or 3 points are given for fewer mistakes. The overall score is the sum of the points on the five mazes, and the fewer the mistakes the greater the total score, the maximum being 21 points.

7. The Japanese version of the WISC (Kodama and Shinagawa, 1953) was administered to the children at their regular schools by the staff of the Department of Psychology, Hiroshima University. The WISC consists of verbal and performance subgroups, each with six parts as listed below with their maximum raw scores. The maze section of the WISC furnished a measure of reliability as it was given both in the survey clinic and in the schools. In general, the Digit Span test was not administered.

The WISC raw scores are converted into scaled scores yielding a verbal IQ, a

Verbal		*Performance*	
General information	30	Picture completion	20
General comprehension	28	Picture arrangement	57
Arithmetic	16	Block design	55
Similarities	28	Object assembly	34
Vocabulary	80	Coding (A or B)	50 or 93
Digit span	26	Mazes	21

performance IQ, and, when these are summed, a full-scale IQ. These IQ's are obtained by comparing each subject's test results with the standard scores for each four-month age group. Within each group the expected full-scale IQ is 100 and the standard deviation is 15.

8. School Grades. Boys and girls in school years 1–4 in the Hiroshima elementary schools are taught Japanese language, social studies, arithmetic, science, music, fine arts, and physical education; beginning with the fifth year, a course on domestic science is added. The student's school performance is scored from 1 to 5 in unit steps. Score "1" is the mark of the lowest 5 per cent, "2" that of the next 20 per cent, "3" of the middle 50 per cent, "4" of the next higher 20 per cent, and "5" of the highest 5 per cent.

Parental migration away from Hiroshima accounted for a 19 per cent loss between the time of establishment of the cohort of children at birth and the time of examination in the Child Health Survey. There is no evidence for systematic difference between the children who left the city and those who stayed and were available for examination as regards parental age, birth rank, or recognized perinatal morbidity.

Table 2 sets out comparative data on the average control child and the average child of single first cousins (whose $F = 1/16$), with reference to the eight classes of behavioral variables. Where appropriate, observations have been standardized to age 120 months.

In each case the mean of the inbred children is significantly depressed compared with the mean of the control children. The observed inbreeding depression is remarkably similar in the psychometric and the school performance variables. Schull and Neel suggest this concordance would seem to imply a high correlation between potential and achievement in these Japanese school children.

The inbreeding effect, computed by regression analysis, is based upon a regression coefficient common to the sexes if these could not be shown to differ. The percentage change with inbreeding varies from 0.9 to 6.7, both extremes falling in the "age when" class. It is interesting to note that the per cent effect of inbreeding on these imprecise measurements is roughly the same as that for the more precise psychometric tests. The per cent change due to inbreeding tends to be greater for the psychometric and school performance than for the neuromuscular scores. The inbreeding effects are greater for the behavioral variables in general than for the measurements of body size which show a small, consistent, and significant inbreeding depression of about 0.5 per cent.

Compared with the controls, the inbred children as a whole come from families of lower socioeconomic status as measured by parental occupation and education, density of persons in the household, and food expenditures per person per month. The effects of socioeconomic status were removed by the use of a socioeconomic score. In boys, about 23 per cent of the apparent inbreeding depression is attributable to socioeconomic variation in the case of neuromuscular tests, 16 per cent for psychometrics, and 25 per cent for school performance. The corresponding percentages for girls are 26, 21, and 38.

TABLE 2. EFFECTS OF INBREEDING ON THE BEHAVIOR OF JAPANESE CHILDREN. (Schull and Neel, 1965)

Characteristic	Sex	Average control child	Average offspring of first cousins	Inbreeding effect	Per cent change with inbreeding	Potential inflation effects of socioeconomic variation	
						Inbreeding effect	Magnitude in %
Age when walked (mos)	M	14.06	14.19*	0.13	0.9	—	—
	F	13.62	14.07**	0.45	3.3	—	—
Age when talked (mos)	M	11.81	12.60**	0.79	6.7	—	—
	F	10.38	10.82**	0.44	4.2	—	—
Dynamometer grip, right (kg)	M	14.04	13.71***	0.29	2.1	0.04	12.1
	F	12.35	12.02**	0.29	2.3	0.04	12.1
Dynamometer grip, left (kg)	M	13.23	13.01**	0.20	1.5	0.02	9.1
	F	11.43	11.20**	0.20	1.7	0.03	13.0
Tapping rate/10 sec., right	M	28.02	27.62**	0.31	1.1	0.09	22.5
	F	26.28	25.85**	0.31	1.2	0.12	27.9
Tapping rate/10 sec., left	M	26.11	25.66**	0.38	1.5	0.07	15.6
	F	24.90	24.42**	0.38	1.5	0.10	20.8
Color trail test score	M	20.54	22.33**	0.68	3.3	1.11	62.0
	F	21.76	23.27**	0.68	3.1	0.83	55.0
Maze tests score	M	17.92	17.50***	0.34	1.9	0.08	19.0
	F	16.98	16.52***	0.34	2.0	0.12	26.1
WISC verbal score	M	58.67	55.34***	2.76	4.7	0.57	17.1
	F	57.01	53.46**	2.76	4.8	0.79	22.3
WISC performance score	M	57.37	54.94**	2.06	3.6	0.37	15.2
	F	55.10	52.52**	2.06	3.7	0.52	20.2
School grade: language	M	3.09	2.95**	0.10	3.2	0.04	28.5
	F	3.28	3.10**	0.10	3.0	0.08	44.4
Social studies	M	3.17	3.04**	0.09	2.8	0.04	30.8
	F	3.14	2.98**	0.09	2.9	0.07	43.8
Mathematics	M	3.21	3.04***	0.13	4.0	0.04	23.5
	F	3.19	2.99**	0.13	4.1	0.07	35.0
Science	M	3.29	3.11**	0.13	4.0	0.04	23.5
	F	3.16	2.95**	0.13	4.1	0.08	38.1
Music	M	2.94	2.78**	0.12	4.1	0.04	25.0
	F	3.34	3.14**	0.12	3.6	0.08	40.0
Fine arts	M	3.09	2.95**	0.10	3.2	0.04	28.6
	F	3.40	3.23**	0.10	2.9	0.07	41.1
Physical education	M	3.28	3.13**	0.13	4.0	0.05	13.3
	F	3.27	3.09**	0.13	4.0	0.05	27.8

** Significant at the 1 per cent level.
* Significant at the 5 per cent level.

ASSORTATIVE MATING

When mates in a breeding population have more phenotypical characters in common than would be expected by chance, that is, by random mating, the pattern is called positive assortative mating; when they have fewer than would be expected by chance, it is called negative assortative mating. Usually the degree of assortment is measured by an association statistic that runs from $+1$ for perfect positive assortative mating to -1 for perfect negative assortment. The product moment coefficient of correlation is used below for continuous traits and the tetrachoric correlation for discrete traits.

We saw that under inbreeding the percentage of heterozygosis in a breeding population is independent of the number of gene loci involved and of their dominance relations. This is not the case under phenotypical assortative mating. The analysis of assortative mating is complicated by the fact that the same phenotype may be due to genotypes of different gene constitution, and the complication increases with the number of loci involved. If dominance is present, a constant correlation $(0 < r_{pp} < 1)$ between the phenotypes of mates $(h^2 = 1)$ in a series of assortative matings results in a changing correlation between their genotypes. Assortative mating acting by itself, like inbreeding, does not change gene frequencies in a breeding population.

First, consider perfect positive assortative mating $(r_{pp} = +1)$ for three phenotypes (2, 1, 0) determined by two autosomal alleles (A, a) without dominance. The frequencies of the three possible kinds of homogamous matings and their offspring are:

Mating			Offspring		
Phenotypes	Genotypes	Frequency	AA	Aa	aa
2×2	AA $\times$ AA	r	r	—	—
1×1	Aa $\times$ Aa	2s	s/2	s	s/2
0×0	aa $\times$ aa	t	—	—	t
Totals		1	r + s/2	s	s/2 + t

In the initial generation, the frequency of gene $A = p_n = r+s$ and that of gene $a = q_n = s + t$; in the offspring generation, their frequencies are

$$p_{n+1} = r + s/2 + s/2 = r + s = p_n,$$
$$q_{n+1} = s/2 + s/2 + t = s + t = q_n,$$

showing the gene frequencies are not changed by this system of assortative mating. The genotype frequencies for the nth generation are

$$r (AA)_n + 2s (Aa)_n + t(aa)_n = 1,$$

but for generation $n + 1$ we have, as given above for the offspring,

$$(AA)_{n+1} = r + s/2$$
$$(AA)_{n+1} = s$$
$$(aa)_{n=1} = s/2 + t.$$

Thus, in general,

$$(AA)_n = (AA)_o + (1/2 - 1/2^n)\,(Aa)_o$$
$$(Aa)_n = 1/2^n\,(Aa)_o$$
$$(aa)_n = (aa)_o + (1/2 - 1/2^n)\,(Aa)_o.$$

Accordingly, the frequency of heterozygotes is halved in each generation and the homozygotes tend toward the following limiting values:

$$\lim (AA)_n = (AA)_o + 1/2\,(Aa)_o = p_o,$$
$$\lim (aa)_n = (aa)_o + 1/2\,(Aa)_o = q_o.$$

This result was first obtained by Jennings (1916). Geppert and Koller (1938) applied the direct method to obtain sequence equations for several other modes of inheritance, including dominance and multiple alleles. If the dominant and recessive alleles are of equal frequency in the population, the percentages of heterozygosis in a system of assortative mating involving two alleles runs:

No dominance: 1/2, 1/4, 1/8, 1/16, 1/32, . . .
Dominance: 1/2, 1/3, 1/4, 1/5, 1/6, ...

The direct method, which considers the outcome of all possible assortative matings, requires too much space to be given in detail here in the case of perfect assortative mating for phenotypes controlled by genes from more than one locus and for all cases where the phenotypical correlation is $0 < r_{pp} < 1$. Wright (1921) carried the work of the direct method to the fourth generation for perfect positive assortative mating for five phenotypes (4, 3, 2, 1, 0) determined by two pairs of equally frequent autosomal alleles (A, a and B, b) lacking dominance and epistasis and assumed to be in equilibrium before the start of assortative mating: the percentages of heterozygosis in successive generations form the series 1/2, 3/8, 10/32, 17/64; the correlation between gametes produced by an individual form the series 0, 1/3, 1/2, 53/87; and the correlation between uniting gametes form the series 1/2, 2/3, 3/4, and 70/87. Fortunately, the method of path coefficients obtains without difficulty general recurrence equations for the genetical consequences of perfect or imperfect assortative mating involving any number of loci with or without dominance.

Figure 5 is a path diagram showing the phenotypes and genotypes of two mated individuals as determined by genes at three autosomal loci lacking dominance and as correlated through correlations among these loci. The path coefficients are defined:

i = the path measuring the influence of the gene pair at a given locus on the genotype—the sum of the effect of these gene pairs.

a_u = the path from A^1 to A^1A^2, the subscript indicating the path measures the effect of a unit factor—the genes at one locus.

e and h are defined as for inbreeding to measure the degree of environmental and of genetic determination of the phenotype.

The correlations shown in Figure 5 are defined:

r_{pp} = correlation between phenotypes of mated individuals (father-mother), the coefficient of assortative mating based on phenotypical resemblance.

m = the correlation between genotypes of mated individuals.

f = the correlation between uniting gametes, as defined for inbreeding. (Not to be confused with f_u.)

f_u = the correlation between genes of the same set of alleles, in different individuals (such as A^1 and A^3).

g_u = the correlation between two genes of the same set of alleles that separate in gamete formation (for example A^1 and A^2), since these genes united at fertilization in the preceding generation, $g_u = f'_u$.

j_u = Correlation between genes from different loci that act on the same phenotypical character (such as A^1 and B^3). It is assumed $j_u = f_u$.

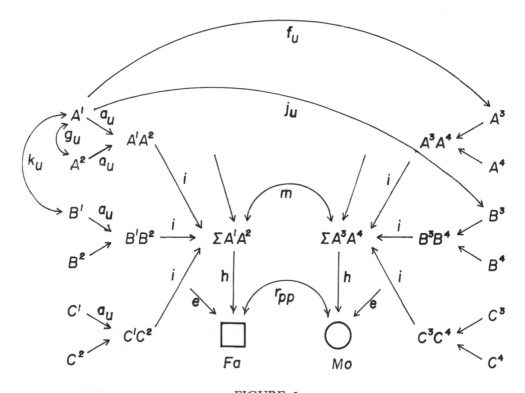

FIGURE 5
Path diagram showing determinants of phenotypical correlations
(after Wright, 1921).

k_u = the correlation between genes of different sets of alleles (such as A and B) in the same individual.

On the basis of these definitions and the path diagram, if we are given n = number of loci (gene pairs), r_{pp} = correlation between the phenotypes of mates (coefficient of assortative mating), and h^2_0 = initial heritability at the start of assortative mating, the following equations may be used to find the effect of any degree of positive or negative phenotypical assortative mating for any number of loci:

$$m = h'^2 r_{pp}$$
$$f_u = (m/2n)[1 + f'_u + 2(n\text{-}1)k_u]$$
$$k_u = (1/2)(f'_u + k'_u)$$
$$H = (1/2)(1 - f_u)$$

Negative values of m are used in the case of negative assortative mating $(0 > r_{pp} > -1)$.

When equilibrium is reached $f_u = (m)/[2n - m(2n - 1)]$, $H = [n(1 - m = [n(1 - m)]/[2n(1 - m) + m]$.

Now we can see that the above equations reach the same results for the effects of assortative mating as those obtained by the direct enumeration of all possible mating pairs. In the case of perfect positive assortative mating for three phenotypes (2, 1, 0) determined by two autosomal alleles (A, a) lacking dominance, we have $n = r_{pp} = h^2_0 = 1$,

$$f_u = (1/2)[1 + f'_u + 0]$$
$$= (1/2)(1 - f_u) = (1/2)H'$$

and one half of the heterozygosis is lost in each generation as found by the direct method.

In the case of perfect positive assortative mating for five phenotypes (4, 3, 2, 1, 0) determined by two pairs of independent autosomal alleles (A, a and B, b) lacking dominance and epistasis, where equilibrium is assumed to have been reached before the beginning of assortative mating, the change in percentage of heterozygosis is given by

$$f_u = (1/4)[1 + f'_u + 2k_u]$$
$$k_u = (1/2)(f'_u + k'_u).$$

Starting with $f_u = k_u = 0$, the above equations yield the following series:

$$k_u = 0, 0, 1/8, 1/4, 23/64, 29/64, \ldots, 1$$
$$f_u = 0, 1/4, 3/8, 30/64, 35/64, 157/256, \ldots, 1$$

Thus, beginning with $H = 1/2$, since $H = (1/2 (1 - f_u)$, we have $H = 1/2, 3/8, 5/16, 17/64, 29/128$, which agree with the fractions obtained by the direct method.

Wright (1921) should be consulted regarding other results of assortative mating, including the effect of dominance, imperfect assortative mating, the consequences of these patterns of mating for the correlation between parent and offspring and between siblings, and the joint effects of inbreeding and assortative mating on the distribution of genotypes in breeding populations. The results for those degrees of positive and negative assortative mating of most anthropological interest are plotted in Figure 6.

In general, population consequences of positive assortative mating are similar to those for inbreeding, since the proportion of heterozygous genotypes is reduced and that of homozygous genotypes increased, in comparison with the proportions expected under random mating. The consequences for multiple gene modes of inheritance differ from those with inbreeding in that fewer of the homozygous genotypes are preserved. For example, if five phenotypes (4, 3, 2, 1, 0) are controlled by two pairs of genes at independent autosomal loci (A, a; B, b) as follows: 4—AABB; 3—AaBB, AABb; 2—AAbb, AaBb, aaBB; 1—Aabb, aaBb; 0—aabb; close inbreeding will result in a genotype distribution with all homozygous classes (AABB, AAbb, aaBB, aabb) preserved, while one of perfect positive assortative mating will preserve only two homozygous classes (AABB, aabb). Negative assortative mating leads to an increase in the proportion of heterozygosis. With perfect assortative mating, no dominance, and complete heritability, equilibrium is not reached until heterozygotes are entirely absent. This triple condition probably never holds for human populations (excluding the uninteresting case of the negative correlation in sex and some sex attributes between mated pairs). In fact, as we will see in the next section, phenotypic correlation above $r_{pp} = 0.5$ are rare in human populations. Thus, in general, assortative mating in human populations may be expected to lead to a condition of genetic equilibrium without fixation of extreme types. A correlation of 0.50 between genotypes would be toward the high end of the range. With one pair of genes involved, heterozygosis is reduced only from 0.500 to 0.333 in an infinite number of generations of assortative mating; with ten pairs, H is reduced to 0.476. Assortative mating, like close inbreeding, leads to increased variability of the population as a whole. For those phenotypes of complete heritability, assortative mating of + 0.50 results in a doubling of the genetic variance of the whole population if an indefinitely large number of gene pairs is involved. The effect is less for a smaller number of gene pairs. With perfect assortative mating, the population reaches equilibrium with completely homozygous genotypes concentrated at the two extremes. Negative assortative mating reduces the genetic variance of the population for phenotypes with high heritability (Wright, 1921).

ASSORTATIVE MATING AND BEHAVIOR

The volume of reliable recorded information on assortative mating in human populations is considerably less than that on inbreeding. And of studies on assortative mating, those concerned with behavioral variables considerably outnumber those on morphological or physiological traits. Table 3 gives an unsystematic

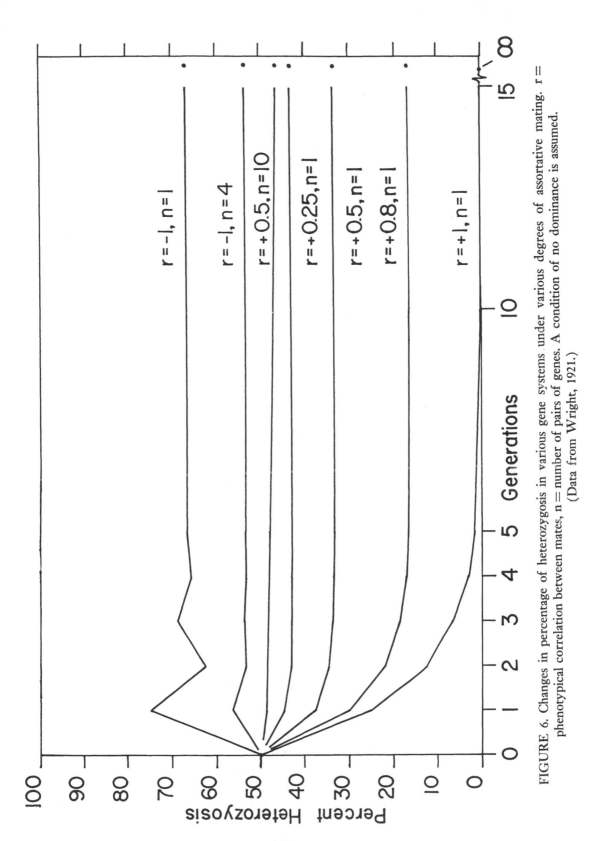

FIGURE 6. Changes in percentage of heterozygosis in various gene systems under various degrees of assortative mating. r = phenotypical correlation between mates, n = number of pairs of genes. A condition of no dominance is assumed. (Data from Wright, 1921.)

TABLE 3

ASSORTATIVE MATING FOR INTELLIGENCE TEST SCORES, PERSONALITY RATINGS,
AND MISCELLANEOUS CHARACTERISTICS.

(Tetrachoric correlations are shown without sigmas.)

Item	Source	N pairs	r
Intelligence scores			
Stanford-Binet	Burks, 1928	174	.47 ± .04
Otis	Freeman *et al.*, 1928	150	.49 ± .04
Army Alpha	Jones, 1928	105	.60 ± .04
Progressive Matrices	Halperin, 1946	324	.76
Various tests	Smith, 1941	433	.19 ± .03
Vocabulary	Carter, 1932	108	.21 ± .06
Arithmetic	Carter, 1932	108	.03 ± .06
Mental Grade	Penrose, 1933	100	.44
Personality ratings			
Neurotic Tendency	Hoffeditz, 1934	100	.16 ± .07
Neurotic Tendency	Terman and Buttenwieser, 1935	126	.11 ± .06
Neurotic Tendency	Terman and Buttenwieser, 1935	215	.22 ± .04
Neurotic Tendency	Willoughby, 1928	100	.27 ± .05
Self-sufficiency	Hoffeditz, 1934	100	.09 ± .07
Self-sufficiency	Terman and Buttenwieser, 1935	215	.12 ± .04
Self-sufficiency	Terman and Buttenwieser, 1935	126	.02 ± .06
Dominance	Hoffeditz, 1934	100	.15 ± .07
Dominance	Terman and Buttenwieser, 1935	126	.24 ± .06
Dominance	Terman and Buttenwieser, 1935	215	.29 ± .04
Introversion-extroversion	Terman and Buttenwieser, 1935	126	.02 ± .06
Introversion-extroversion	Terman and Buttenwieser, 1935	215	.16 ± .04
Miscellaneous			
Temperament	Burgess and Wallin, 1944	316	.22
Insanity	Goring, 1909	1433	.06
Criminality	Goring, 1909	474	.20

selection of 25 studies based on samples of 100 mated pairs or more. Many, if not most, traits investigated probably have low heritability; for some of the traits the observed correlation probably developed or increased after marriage rather than prior to it. The unweighted mean of the 21 product-moment correlations in this table rounds to 0.25 and the range goes from r-values not significantly different from zero to 0.62 ± .05.

A number of highly capable, statistically oriented biologists, not motivated by any particular political outlook, have expressed concern about the possible decline in intelligence in the general population deduced from the negative correlation between the measured intelligence scores of children and the size of the families they represent. Several large-sample projects have established small, negative correlations (in the range -0.3 to -0.2 and of unquestioned significance) between the number of sibs and the intelligence scores of school-age children in various countries, including France, New Zealand, Scotland, and the United States.

The decline argument goes this way: Test results are obtained for a sample of school children, say, of age 11. It is assumed that the tested child's score may be taken as an estimated score of his untested siblings. A mean score is calculated for children of each sibship size. The mean score of the untested *parental generation* is estimated from the mean scores of each sibship size class weighted by the number of subjects in that class.

The mean score of the *offspring generation* is estimated by weighting the number of subjects in a sibship size class by the size of the sibship. Given the negative correlation between intelligence scores and sibship size, the estimated mean score of the offspring generation is necessarily lower than the estimated mean score of the parental generation and the difference is taken as a measure of the decline in average intelligence scores in one generation (Giles-Bernardelli, 1950).

The argument for a decline in the average level in intelligence is accepted with little question by some workers despite the fact that test scores on school children are observed to be higher than those of about a decade earlier.

Higgins, Reed, and Reed (1962) and Bajema (1963) identified a serious defect in the decline argument insofar as it is based on the negative correlation between intelligence scores and sibship size ascertained from school children. That part of the population that does not reproduce (as much as 20 per cent in Bajema's Michigan study) is not represented in a sample drawn from school children. The relationship between the intelligence of an individual and the size of the family from which he comes is biased in a negative direction because as intelligence decreases the probability of childlessness increases, as a result of either differential marriage rates or differential childbearing rates.

Using a life table analysis to calculate the intrinsic rate of natural increase, Bajema found the Kalamazoo, Michigan, population was either very close to equilibrium with respect to genetic factors favoring high intelligence or, more likely, had experienced a recent slight increase in the frequency of such factors, despite the fact that the observed correlation between intelligence score and completed family size is -0.26 (significant at the 1 per cent level).

Penrose (1949) produced a hypothetical model demonstrating that perfect positive assortative mating for intelligence (assumed to depend on a single pair of autosomal alleles) leads to an equilibrium with respect to the distribution of the genotypes concerned with intellectual ability, if the birth rate of the heterozygous inferior group is twice that of the homozygous superior group and if the inferior group just replaces their numbers, because one quarter of their offspring go into the superior group, one half stay in the inferior group, and one quarter are subnormal and infertile.

An obvious improvement in the design of studies on the supposed population decline in intelligence would correlate the tested intelligence scores of *both* parents with the number of their children. I will conclude this report by presenting the results of two cognitive tests carried out in a study of assortative mating in the Ann Arbor population (Spuhler, 1962), together with heritability estimates for these tests based on a one-egg and two-egg twin sample from the area centering about Ann Arbor.

The first test is the Progressive Matrices (Raven, 1946), which is described as "a non-verbal test of a person's capacity at the time of the test to apprehend figures presented for his perception, see relations between them, and conceive the correlative figures completing the system of relations presented." When used without a time limit, the Progressive Matrices constitute a "test of intellectual capacity." The present analysis of the test scores is based on the total number correct with a possible range from 0 to 60.

The second test is the "Verbal Meaning" part of the Chicago Tests of Primary Mental Abilities (Thurstone and Thurstone, 1941). The present analysis is restricted to the scores on the "Sentences" part of the test, which involves selection of one of four words that best completes the meaning of each of forty sentences. Three minutes were allowed for practice and five minutes for the test. The maximum score is 40. The present analysis considers (a) the total number of right answers, and (b) the proportion of right answers.

Table 4 indicates the observed correlations between spouses for the three test results are all significant beyond the 1 per cent level.

TABLE 4

ASSORTATIVE MATING FOR THREE SCORES ON TWO INTELLIGENCE TESTS.
(Spuhler, 1962)

Statistic	Progressive matrices	Chicago Verbal	
	Total right	Total right	Proportion right
No. of couples	180	151	148
Father's mean	35.60	22.21	0.93
Mother's mean	34.99	20.86	0.91
Correlation	+ 0.399**	+ 0.305**	+ 0.732**

** Significant at the 1 per cent level.

The heritability of these test scores is not known for the general population of Ann Arbor but their general magnitude may be inferred from results based on the Ann Arbor twin sample:

	Twin pairs	h^2	P
Progressive Matrices, total right	76	0.94	(P < .05)
Sentences, total right	82	0.45	(P < .05)
Sentences, proportion right	82	0.40	(.05 < P < .1)

The correlation between fertility and intelligence scores of the married couples in our sample may be obtained by use of a fertility index. Many of these couples had not completed their reproductive period. A square root transformation of months of exposure to pregnancy plotted against the number of live-born children showed a strong, approximately linear relationship between exposure and fertility. From this relationship a score was obtained giving the difference between observed and expected fertility for each couple. The range of this fertility score was from -2.57 to $+3.49$, with a mean of $+0.0003$ and a standard deviation of 1.06. The distribution of the fertility scores was found to be satisfactorily close to a normal distribution. The correlation between the test scores and the fertility indices are given in Table 5.

TABLE 5

CORRELATION OF INTELLIGENCE TEST SCORES AND FERTILITY INDICES.

(Spuhler, 1962)

Score of	Progressive matrices Total right		Chicago Verbal Total right		Proportional right	
	N	r	N	r	N	r
Mother	180	$+0.148*$	151	$+0.128$	151	$+0.234*$
Father	180	$+0.032$	151	-0.010	151	$+0.038$

* Significant at the 5 per cent level.

The correlations are positive and significant at the 5 per cent level between mother's fertility index and her score on two of the three tests. The correlations are not significantly different from zero for father's fertility index and test scores.

Just where the population of Ann Arbor falls in the distribution of all possible samples from human populations is, of course, unknown. With regard to the evidence from our small sample, we may conclude that there is significant positive assortative mating for intelligence as estimated by scores on three cognitive tests; that one of the tests show strong, another moderate, and the third no heritability; and that there is a positive correlation between tested intelligence of mothers and their reproductive performance.

BIBLIOGRAPHY

BAJEMA, C. J.
1963. Estimation of the direction and intensity of natural selection in relation to human intelligence by means of the intrinsic rate of natural increase. *Eugen. Quart.*, 10:175–187.

BELL, J.
1941. A determination of the consanguinity rate in the general hospital population of England and Wales. *Ann. Eugen. Lond.*, 10:370–391.

BONNÉ, B.
1963. The Samaritans: a demographic study. *Hum. Biol.*, 35:61–89.

BUNAK, V. V.
1965. Der Verwandtschaftsgrad der Bevölkerung kleiner ländlicher Gemeinden. *Acta Genet. Med. et Gemmellog.*, 14:174–181.

BURGESS, E. W., and P. WALLIN
1944. Homogamy in personality characteristics. *J. Abn. and Soc. Psychol.*, 39:475–481.

BURKS, B. S.
1928. The relative influence of nature and nurture upon mental development. *Nat. Soc. Stud. Educ.*, 27 (1):219–321.

CARTER, H. D.
1932. Family resemblances in verbal and numerical abilities. *Genet. Psychol. Monog.*, 12:1–104.

EAST., E. M., and D. F. JONES
1919. *Inbreeding and outbreeding.* Philadelphia: J. B. Lippincott.

FALCONER, D. S.
1960. *Introduction to quantitative genetics.* Edinburgh: Oliver and Boyd.

FORD, C. S.
1945. A comparative study of human reproduction. *Yale University Pub. Anthrop.* No. 3.

FREEMAN, F. N., J. HOLZINGER, and B. C. MITCHELL
1928. The influence of the environment on the intelligence, school achievement and conduct of foster children. *Yearb. Nat. Soc. Stud. Educ.*, 27 (1):103–217.

GEPPERT, H., and S. KOLLER
1938. *Erbmathematik.* Leipzig: Meyer Verlag.

GILES-BERNARDELLI, B. M.
1950. The decline of intelligence in New Zealand. *Pop. Studies*, 4:200–208.

GORING, C. H.
1909. Studies in national deterioration. *Draper's Company Research Memoir*, 5. London: Dulau.

HALPERIN, S. L.
1946. Human heredity and mental deficiency. *Amer. J. Ment. Def.*, 51:153–163.

HIGGINS, J., E. REED, and S. REED
1962. Intelligence and family size: a paradox resolved. *Eugen. Quart.*, 9:84–90.

HOFFEDITZ, E. L.
1934. Family resemblance in personality traits. *J. Soc. Psychol.*, 5:214–227.

HOLMES, S. J.
1924. A bibliography of eugenics. *Univ. Calif. Pub. Zool.*, Vol. 25.

HUTH, A. H.
1875. *The marriage of near kin.* London: Longmans, Green.

JENNINGS, H. S.
1916. The numerical results of diverse systems of breeding. *Genetics*, 1:53–89.

JONES, H. E.
1928. A first study of parent-child resemblance in intelligence. *Yearb. Nat. Soc. Stud. Educ.*, 27 (1):61–72.

KEMPTHORNE, O.
1957. *An introduction to genetic statistics.* New York: John Wiley.

KODAMA, H., and F. SHINAGAWA
1953. *WISC Chinō shindan kensahō* [*The WISC intelligence test*]. Tokyo: Nihon Bunka Kagakusha.

KUDO, A.
1962. A method for calculating the inbreeding coefficient. *Am. J. Human Genet.*, 14:426–432.

KUDO, A., and K. SAKAGUCHI
1963. A method for calculating the inbreeding coefficient. II. Sex-linked genes. *Am. J. Human Genet.*, 15:476–480.

LINTON, R.
1955. *The tree of culture.* New York: Alfred A. Knopf.

LORIMER, F.
1954. *Culture and human fertility.* Paris: Unesco.

MORTON, N. E., C. S. CHUNG, and M. P. MI
1967. Genetics of interracial crosses in Hawaii. *Monographs in Human Genetics,* Vol. 3. Basel: S. Karger.

MOURANT, A. E.
1954. *The distribution of the human blood groups.* Oxford: Blackwell.

NEEL, J. V., M. KODANI, R. BREWER, and R. C. ANDERSON
1949. The incidence of consanguineous matings in Japan with remarks on the estimation of comparative gene frequencies and the expected rate of appearance of induced recessive mutations. *Am. J. Human Genet.*, 1:156–178.

PEARL, R.
1915. *Modes of research in genetics.* New York: Macmillan.

PENROSE, L. S.
1933. A study in the inheritance of intelligence. The analysis of 100 families containing subcultural mental defectives. *Brit. J. Psychol., Gen. Sec.,* 24:1–19.

1949. *The biology of mental defect.* New York: Grune and Stratton.

RASMUSON, M.
1961. *Genetics on the population level.* Stockholm: Svenska Bokförlaget.

RAVEN, J. C.
1946. *Progressive matrices.* London: H. K. Lewis.

SCHULL, W. J., and J. V. NEEL
1962. The Child Health Survey: a genetic study in Japan. In *The use of vital and health statistics for genetic and radiation studies.* New York: United Nations.

1965. *The effects of inbreeding on Japanese children.* New York: Harper and Row.

SMITH, M.
1941. Similarities of marriage partners in intelligence. *Amer. Sociol. Rev.* 6:697–701.

SPUHLER, J. N.
1962. Empirical studies on quantitative human genetics. In *The use of vital and health statistics for genetic and radiation studies.* New York: United Nations.

SPUHLER, J. N., and C. KLUCKHOHN
1953. Inbreeding coefficients of the Ramah Navaho population. *Human Biol.,* 25: 296–317.

SUTTER, J., and L. TABAH
1948. Fréquence et répartition des marriages consanguins en France. *Population,* 4:607–630.

TERMAN, L. M., and P. BUTTENWIESER
1935. Personality factors in marital compatibility. *J. Soc. Psychol.*, 6:143–171, 267–289.

THURSTONE, L. L., and T. G. THURSTONE
1941. *The Chicago tests of primary mental abilities*, *V*. Chicago: Science Research Associates.

WESTERMARCK, E. A.
1903. *The history of human marriage*. (3d. ed.) London: Macmillan.

WILLOUGHBY, R. R.
1928. Family similarities in mental test abilities. *Yearb. Nat. Soc. Stud. Educ.*, 27 (1):55–59.

WRIGHT, S.
1921. Systems of mating. *Genetics*, 6:111–178.

1933. Inbreeding and recombination. *Proc. Nat. Acad. Sci.*, 19:420–433.

1950. Discussion on population genetics and radiation. *J. Cell. and Comp. Physiol.*, 35 (Suppl. 1):187–210.

1951. The genetical structure of populations. *Ann. Eugen. Lond.*, 15:323–354.

WRIGHT, S., and H. C. McPHEE
1925. An approximate method of calculating coefficients of inbreeding and relationship from livestock pedigrees. *J. Agric. Res.*, 31:377–384.

BEHAVIORAL CONSEQUENCES OF
GENETIC DIFFERENCES IN MAN: A SUMMARY

ERNST W. CASPARI

THE PRESENT CONFERENCE has as its topic "behavioral consequences of genetic differences in man." This title alone implies that a large variety of different types of information has to be brought to bear on a single set of problems. On the one hand, we have dealt with the description and analysis of behavior. It includes problems of the proper description of behavioral phenotypes; their classification and appropriate analysis; the inheritance, mechanisms, and development of the characters studied; and their function, particularly with respect to social organization.

The second aspect treated extensively is the genetic basis of human diversity. In this area of discussion, much emphasis has been given to polygenic inheritance and the methods for its analysis, the phenomenon of genetic polymorphism particularly in human populations, the analysis of mating systems, and the interaction of genetic and environmental influences on the determination of behavioral phenotypes.

Finally, it appeared to be necessary to study the problem from an evolutionary point of view. The two main classical methods of phylogenetic investigation—comparative study of related organisms and study of the remains of prehumans and early humans by paleontological and archeological methods—as well as more modern methods using gene frequencies in populations have been applied to this problem. It remains to summarize the main conclusions from these diverse lines of investigation, and to emphasize their bearing on the primary problem of this conference.

DESCRIPTION AND CLASSIFICATION OF BEHAVIOR

The problems involved in the description of behavior have been most thoroughly discussed by Hirsch and by McClearn. Hirsch defines behavior as that part of the activities of an organism that we choose for study. A behavior, as defined in this way, can then be studied in terms of mechanism and function. Obviously, the principles used in the choice of a behavioral phenotype are an important aspect which has been treated explicitly by McClearn. The behavior chosen should be operationally defined by a psychological test having the following properties: (1) it should be reliable, that is, reproducible for an individual; (2) there should be a certain amount of variation for this test in the population studied; and (3) the tests should be appropriate in the sense that a balance must

be struck between two opposing tendencies—more global tests may be more meaningful in interpreting the function of a behavioral phenotype in its natural setting,
but narrower tests are more likely to give consistent and reproducible results and
also to lead to the establishment of "natural units." In most cases, a test should
be minimally subject to modification by environmental factors, unless the investigator is particularly interested in studying heredity-environment interactions.

Hirsch points out that after a behavioral phenotype has been chosen and operationally defined, it has to be described in terms of function and mechanism. Such
descriptions will lead to a taxonomy of behavioral characters and will permit
analysis by breaking it down into components. If the analysis proceeds successfully, it will result in the isolation of natural units of behavior.

The procedure of classification inherent in Hirsch's proposal of a logical analysis of behavior has been subjected to a critical and searching discussion in the
paper by Dobzhansky. He emphasizes that classification is a primary and necessary approach to the study of any set of natural phenomena by man. It has been
stated repeatedly, not only by Dobzhansky, but also by other participants, particularly Spuhler, that human language itself uses classification as its main approach
to the ordering of the environment.

A correct classification leads to predictions that can be verified by further observations. The main danger of classification consists in the fact that the categories
resulting from it may be taken to constitute realities. This typological approach
has led to a great number of difficulties not only in scientific research, but also
in applications to everyday life. Genetics has led to an alternate approach, termed
the population or existential approach, which accepts the genetic differences of
individuals in a population as a fact and describes the population in dynamic
terms. Nevertheless, the temptation to continue to use typological concepts remains large. Kalmus points out that stable equilibria in populations will frequently
mislead an investigator into thinking in typological concepts. In the same way
the existence of a relatively invariant wild type and of a population mean for
variant characters carries the danger of typological thinking. The constitutional
types described by Kretschmer and Sheldon, as well as the concept of an "optimal genotype," are striking examples of typology that persist in the present scientific literature.

Questions have been raised concerning the meaning of "natural unit of behavior" proposed by Hirsch. Hirsch expects that such units will appear when
a behavioral character is investigated by appropriate techniques. But Dobzhansky
and Thoday feel that it is an imprecise term, and Dobzhansky proposes to use
the term "convenient unit" instead. This does not seem to cover Hirsch's concept, however, since he feels that the units revealed by the classification of behavioral phenotypes represent real phenomena. Dobzhansky emphasizes that the
only units that are transmitted biologically are individual genes or, at most, co-
adaptive gene complexes held together by inversions or close linkage, but not
genotypes. It must therefore be concluded that natural units, if they are encountered in the taxonomy of behavior, must be either due to single genes or based

on gene complexes which, in the evolutionary history of an organism, have become coadapted by selection.

THE GENETIC DETERMINATION OF BEHAVIOR

Numerous examples were cited, particularly by Böök and by Kalmus, in which behavioral characters depend on alleles of a single gene. Böök emphasized pathological characters, particularly mental deficiency, while Kalmus quoted examples for the isolation of single genes that are regarded as nonpathological and affect sensory perception. On the other hand, much discussion has been devoted to polygenic inheritance, since most characters as defined by psychological tests have turned out to be polygenic in their inheritance. Böök reported that many, though not all, chromosome aberrations found in man have some influence on intelligence, and this fact points to the conclusions that the number of loci affecting intelligence in the human genome must be large and widely distributed among the chromosomes and that intelligence is affected not only by differences in allelic states of these loci, but by their quantitative balance as well.

Thoday's paper indicates that the difference between polygenic systems and individual genes is not absolute. By an ingenious technique he has been able to demonstrate that 88 per cent of a strain difference in a quantitative character can be accounted for by the action of two alleles, each at five identifiable loci, and their interactions with each other. Each of these loci has a clearly defined phenotypic effect on the character in question. It appears likely that the same method in principle can be applied to any character that appears to be determined by polygenes, provided that the chromosomes of the organism in question are well mapped and provide good marker genes. This limits the number of animal species whose behavior could be studied in this way to *Drosophila* and the mouse, and possibly the chicken and the silkworm.

It is important to recognize that the difference between unigenic and polygenic determination of a character depends on the method used to investigate variation. If selection is practiced for a morphological, behavioral, or physiological character, the outcome will usually be a change in the character that, on crossing, turns out to depend on several genes. The same result is to be anticipated, if strains established by inbreeding are tested for differences in any quantitative character, and the inheritance of such a character is investigated. If an investigator working on animals wants to obtain differences dependent on single genes, he has to wait for a spontaneous mutation to occur, or he can speed up the process by the use of mutagenic agents and appropriate selection techniques.

The same principle applies *mutatis mutandis* to man. Most of the simple genic differences encountered in man are rare pathological conditions that are kept in the population by a mutation-selection equilibrium. They must be pathological, because selection against the gene must be strong; and they will be rare, because the mutation rate for individual genes is low. In other words, these rare pathological conditions in man are equivalent to the mutations we obtain in animals by patient waiting. A different type of unigenic polymorphism encountered in human populations concerns characters that reflect closely the constitution of the genetic DNA, in other words, the structure of proteins and of antigens. Differences in human populations in characters other than these two groups are usually polygenic and frequently turn out to be the result of complex gene-environment interactions.

A large part of the conference has therefore been devoted to the discussion of methods that may be used to analyze the polygenic systems involved in the determination of behavior. Admittedly, the situation is complex and hard to analyze. Nevertheless, no emphasis should be placed on the complexity of the situations, since this might discourage progress in the area of research. Rather, it should be stated that Thoday, Kalmus, L. Guttman, Cavalli, McClearn, Vandenberg, and Böök have all discussed useful methods that have already given valuable insights into the polygenic systems affecting variations in the behavior of man.

The most general method described in the greatest detail is the multivariate analysis proposed by L. Guttman. It depends in principle on a covariance analysis of the phenotypes of parents and offspring, including a large number of variables. A correlation picture that appears at first sight very complex or even meaningless may become manageable after the introduction of one or more additional variables. The resulting pattern of correlations will establish relationships that are invariant under different environmental conditions and therefore give evidence of genetic determination. Guttman explained his method by presenting data on the school performance of Israeli children of different origin in six different subjects. In addition, the analysis by Kalmus of his data on the ability to taste bitter substances, Thoday's analysis of a polygenic system by means of marker genes, and the consanguinity studies reported by Cavalli, as well as his evolutionary investigations constitute special instances of the use of the method of multivariate analysis.

A different approach to the study of the determination of polygenic inheritance in man is the use of human isolates presented by Böök. In the study of schizophrenia in an isolated population, it is possible to trace back the condition to a very low number of afflicted ancestors; and therefore the presumption that the condition appearing in this isolate has an identical genetic basis appears legitimate. Caution must be exerted, however, in applying data obtained on an isolate to other populations, since phenotypically identical conditions may have different genotypic determinants.

Mechanism and Development of Behavioral Phenotypes

It is generally agreed that the anatomical substrates for the behavioral characters are mainly localized in the brain, though sensory (Kalmus) and motor structures play a role. In addition, the hormones produced by some of the endocrine glands, such as adrenals, thyroid, and gonads, play a major role in the determination of behavior at the phenotypic level. Genes are primarily agents that determine the properties of individual cells, and development is the mechanism by which these cellular properties are amplified into organismic characters. The problem of genic action as it applies to behavioral characters presupposes a knowledge of structure and function of the organs involved, particularly the central nervous system, the hormonal system, and their interrelations. Hamburg, in discussing this subject, points out that, from a functional point of view, the nervous and the endocrine systems constitute *one* system concerned with short-term adaptation to environmental changes. The brain is a highly complex organ which develops primarily without the intervention of environmental action; the differentiation of the different parts of the brain and their interconnection by fiber tracts are early embryological processes. These interconnections determine the primary functional relations of the parts of the brain. Hamburg placed particular emphasis on the hypothalamus since this structure is connected with two main higher centers of the brain, which affect it in different directions and react particularly on stimuli involving emotions. The hypothalamus controls the pituitary gland, which in turn affects the action of several hormone glands, and determines in this way the hormone level in the blood. A homeostatic feedback reaction of the hormone level of the blood on the hypothalamus may occur. The "hormonostat" in the hypothalamus seems to appear early in development. Genetic influences on this hormonostat can be most easily analyzed from the point of view of the hormones, because the hormones themselves are chemically well-identified substances, and sensitive methods for their determination in blood and urine are available. The environmental conditions that lead to their secretion and their persistence in the blood are, therefore, well investigated. Finally, their biosynthesis has been studied, and the consequences of genetic blocks, partial or complete, in their synthesis on the behavior of affected organisms have been investigated. These behavioral changes can be described in the usual terms of physiological genetics, in other words, as results of lack of the product of a chemical reaction, accumulation of a precursor, or further metabolism of the precursor along an alternate pathway.

The analysis of genetic effects on the central parts of the hormonostat are more difficult to analyze, since the primary genetic effects on metabolism in these cells are not known. Hamburg points out, however, that differences in the sensitivity of the hypothalamic part of the hormonostat between mouse strains are present at an early age, and McClearn reports confirmatory evidence that stimulation of pregnant mothers from different mouse strains may have different effects, not only quantitatively, but even going in opposite directions, in the offspring.

Investigations on the development of behavioral activities were reported by Papoušek. In animals, brain damage at early stages can be repaired, at least in a functional sense, if it occurs before a critical period characteristic for different behavioral characters. In human beings, conditioning is possible very shortly after birth. Neonates can be conditioned to react to stimuli by turning their heads, and very early can distinguish between the right side and the left side. It is important to realize that the limitation to the conditioning of early infants resides in the development of their motor abilities. Conditioning goes on at a stage of development when the central nervous system is by no means fully differentiated.

Particularly the myelinization of the nerve fibers in the brain occurs only after birth.

Individual differences in conditioning show themselves in Papoušek's material by the strength of the conditioned response. Hamburg points out that the children in Papoušek's study were isolated at such an early age that individual differences cannot be ascribed to the influence of social conditions. Both the strength of the conditioned reaction and the time necessary for conditioning change as the child grows older. Individual differences are also found in older children, but the relative position of a child in the whole group may change with time. Individual differences are, therefore, not limited to any one point in time, but may represent individual differences in the rate of behavioral development.

In the preceding paragraphs, heredity-environment interactions have been repeatedly mentioned. In Papoušek's materials, individual differences of a quantitative kind were found that may be presumed to be at least in part genetically determined, since the children had not yet been exposed to different social environments. But Hamburg's data present a more complete model for the interaction of genetic background and environmental conditions.

Genetic control is indicated for at least two components of the hormonostat. One factor is a property of the hypothalamus that may express itself as an innate degree of sensitivity to stimuli provided by the higher centers of the brain. In addition, Hamburg points out that individuals differ in the consistency of their reactions at different times. Some people show high variability in time, while other people react in a more uniform manner. Thoday points out that variability —the ability to vary—must be strictly distinguished from the variation observed. The former property is, in my view, frequently overlooked in morphological studies since the morphological development of animals is well canalized. Variability plays, however, a large role in the analysis of the behavior of animals and of the morphological characters of plants.

The other component of the hormonostat in which gene-environment interactions can be clearly demonstrated is the hormonal control. These individual differences in reaction are reasonably well understood, and particular genetic blocks can be shown to lead to pathological reactions to emotional stimuli.

In human beings, the heredity-environment question centers around the question of the influence of the social environment on the behavior of a human being. There is no doubt that learning, and particularly early learning, has a profound influence on the behavior of human beings. Furthermore, since parents and offspring are situated in closely related social environments, a behavioral correlation between parents and offspring may indicate a genetic influence, an influence of the social environment, or both. The emphasis put on one or the other of these factors depends to a certain degree on the mental attitudes of the investigator. On the other hand, the relation of these two components in the determination of human behavior appears to be an important scientific question that is one of the main reasons for regarding behavior genetics as a separate discipline.

One approach to this problem is the study of behavioral characters that may be presumed to be relatively independent of the social environment. Examples are the early conditioning of infants reported by Papoušek and McClearn's investigations on subitizing. L. Guttman's multivariate analysis may achieve the same goal for more complex behavioral characters.

The classical method for studying the relation of genetic and environmental factors are twin studies, which have been discussed by Vandenberg and by Böök. Vandenberg stressed the severe limitations of the twin method for the study of heredity. At most it can give some indication that genetic factors are involved in the determination of a character, but it is unable to give any information on the mode of inheritance and the genetic structure of a character. This difficulty is inherent in the twin method, since it consists in the comparison of genetically identical organisms. Comparison of identical organisms, however, is a method for the study of environmental influences, not of genes. Particular emphasis should therefore be given to discordant monozygotic twins, since they could be expected to give information on environmental factors affecting a character in a specific genetic constellation. In this connection, Böök quoted studies on monozygotic twins discordant with respect to schizophrenia, in which a particular psychological trauma could be identified as responsible for the onset of the disease in the affected twin. Vandenberg reported on the systematic use of monozygotic twins in educational studies, the two twins serving as controls for each other. The results indicate a strong genotype-environment interaction insofar as the reaction of different pairs of twins to two educational methods may be quantitatively different or even tend in opposite directions.

A particular limitation of twin studies is that the twins themselves form a rather peculiar type of social system, one twin being part of the environment of the other. Hamburg pointed out that different pairs of monozygotic twins react differently on their situation as twins.

As has been emphasized repeatedly, most strongly by Hirsch and by Hamburg, behavior is a method of organisms to react to their environment, physical as well as social. The problem of whether a certain behavioral character is determined by genes or by the environment, or even a partitioning of the variance into genetic and environmental variance, is therefore misleading from the point of view of understanding behavior. Rather, the question, particularly with respect to human psychology, is how genetic individuality, which is a demonstrated fact, will express itself under the influence of diverse social and environmental conditions.

EVOLUTION OF HUMAN BEHAVIOR

The evolution of human behavior has been discussed by DeVore from a comparative point of view and by Kurth by means of paleontological methods. Both authors point out that in many respects ancestral human behavior corresponded to typical primate behavior. All primates live in relatively small groups. Change of individuals from one group to another occurs, but it is rare. Generally speaking, the primate society consists, therefore, of an ingroup that is, at the same time, a mating association and that distinguishes itself from outgroups of the same species. Inside the group, individuals can distinguish other individuals from each other and react differently to different individuals. The rearing of infants is a group activity, and the two sexes have different roles in the activities of the group. Most primate groups have ranges, but no defended territories. A baboon troop has no specified base, but Kurth points out that orangs, though now restricted to arboreal life, may have had relatively stable home bases in the Pleistocene. DeVore suggests that primitive prehominids had fixed home bases, even though the so-called living floors were probably used for butchering the prey.

In baboons, the capacity for learning is large and inherited, and it is used for the interpretation of signals on which group cooperation is based. DeVore regards the specific vocalizations of baboons as innate, but the responses to different sounds are learned. Therefore, in particular baboon troops, there may arise a tradition of behavior based on learning of the meaning of signals by the young. While the behavior of baboons in general appears species specific, subtle differences in the social behavior of different troops can be recognized.

Man is distinct, according to DeVore, for his meat diet and for his use and making of tools, as well as bipedal locomotion, his dentition, and the size of his brain. To these characters, Spuhler adds the extensive use of symbols, particularly in language, and Kurth insists on the particular character of perseverance in his activities. Meat is eaten by other primates occasionally, and Kurth quotes observations of chimpanzee groups that hunt antelopes. Nevertheless, meat seems to have played a larger role in the nutrition of man than in any other primate. As DeVore points out, this meat diet implies the use of tools, since Australopithecines could not even have survived as scavengers without tools. While tool use and even toolmaking is known to occur in some higher primates, it is characteristic of the perseverance of man that he continues to make the same tools over and over again. Kurth points out, furthermore, that for a large period of time, perhaps two million years, the tools did not undergo any great changes, while at the same time man changed considerably from a morphological point of view. In the last 100,000 years, this relation between morphological and cultural change has been reversed.

Cavalli has presented an analysis of human evolutionary relationships in the more recent past by the ingenious use of a method of multivariate analysis. The characters investigated consist of such phenotypes whose genetic basis is well understood, and for which most human populations are reasonably polymorphic, mainly blood antigens. Assuming that the variations found today are originally due to drift, the data support the assumption that the present human populations became separated from each other about 100,000 years ago, and that the center of diversification was Southeast Asia. The estimates of human population sizes for different times given by Kurth and the archeological evidence tend to support Cavalli's conclusion.

MECHANISMS OF GENETIC POLYMORPHISM

The mechanisms responsible for the persistence of genetic polymorphism in populations have been discussed by Dobzhansky and by Thoday. Dobzhansky emphasizes the stable polymorphism produced by heterozygote superiority, but in addition he points to other mechanisms, such as diversifying selection, dependence of mating frequency on the frequency of the genotype (Ehrman), and meiotic drive. Thoday discusses more intensively particular systems of selections that give rise to polymorphism.

The mating systems in human societies have been discussed by Spuhler, with respect to American whites and Navajo Indians, and by Cavalli, with respect to Italian communities. In both investigations, the emphasis was on consanguineous marriages and assortative mating. Spuhler points out that while in more primitive societies consanguineous marriages were more frequent, assortative matings, particularly with respect to social level, are more important in the more complex modern societies. Cavalli particularly emphasized the social customs leading to assortative mating in his discussion of the population genetic consequences of the "commère" system of marriages, and of migration. It appears doubtful, however, how far the mating systems discussed contribute to the maintenance of genetic polymorphism in human populations.

The Importance of Genetic Polymorphism in Human Societies

The amount of genetic polymorphism in human populations is very high. Keiter reports evidence that, in the variation of the nose in European populations, at least 26 different traits, which vary independently and show heritability, can be distinguished. These traits have been distinguished from each other by anthropometric methods, and constitute, therefore, an example of "natural units" in the sense of Hirsch. Their genetic basis has not yet been investigated.

Genetic polymorphism for physical characters is highly important for the behavior of organisms in a social context. DeVore points out that in a baboon troup, the individuals recognize each other by visual cues arising from differences in physical characters, and that such characters, physical as well as behavioral, determine the dominance position and the function of an individual in the group. Thoday and I both point out that this principle must be applied to all vertebrates that show social organization. In a chicken flock, the individual occupies a specific position in the peck order; and after it has been removed from the flock for some time and then reintroduced, it will immediately occupy its proper place in the order of dominance. This implies that the other chickens recognize the individual, a rather remarkable feat of memory for an animal not particularly noted for its intelligence, and it is known that chickens recognize each other by visual cues.

Identification of individuals is essential in human societies for the proper functioning of the social organization. Furthermore, it is common knowledge that this identification depends primarily on visual cues, particularly affecting characters of the face. In other words, genetic polymorphism of the type analyzed by Keiter constitutes the basis of human social organization. The importance of this fact can be best seen when genetically very similar individuals, such as monozygotic twins, are present in a social setting, which frequently leads to difficulties. While polymorphism affecting the face is most frequently used for the identification of individuals in societies, polymorphism affecting other characters may be just as effective. Authorities have shown a preference for using the polymorphism in the pattern of the finger ridges for this purpose, probably because they are two-dimensional and can be more easily filed.

In addition to its role in the identification of individuals, genetically determined polymorphism in the structure of the face seems to be important for the social role an organism will assume. Keiter points out that, on the basis of visual cues arising from the face, human observers develop specific expectations regarding the behavior of a particular individual. There seems to be good agreement, at least among observers sharing the same cultural background, on the expectations derived from seeing a particular combination of characters. In other words, the reactions of other individuals to a person will depend, to a large extent, on the characters of his face; these are therefore important for his social role.

The importance of genetic polymorphism becomes more complex when communication within a society is considered. Baboons, as well as humans, communicate with each other by auditory and visual cues. In baboons, both types carry more or less equal weight, while in man communication by sound is by far the most important. DeVore points out that the cues themselves are probably innate in baboons, but the reaction to the cues is learned. In humans, at most the expression of strong emotions may be regarded as innate. Otherwise, the ability to learn to communicate has a genetic basis, but the particular language used is exclusively dependent on learning. Spuhler points out that strong reliance on learning in behavior concerned with communication inside a group is essential if the group consists of genetically different individuals. A genetic basis of communication would be workable only in a social system, such as that of the honey bee, where the individuals in a hive are all closely related to each other genetically and where genetic polymorphism is therefore very low.

A further question of great interest—the social importance of genetic polymorphism with respect to behavioral characters and abilities—has been only briefly discussed, probably because little material bearing on the question is available except for pathological characters, which represent part of the genetic load of human populations. L. Guttman's multivariate analysis of school performance indicates that genetic variability with respect to such characters exists, and in a number of investigations evidence has been presented for the genetic determination of specific intellectual abilities. In a society such as the human, which has shown an increasing tendency toward division and differentiation of functions during the last 6,000 years, the genetic basis of the ability to assume particular roles would seem to be most important. At the present time, however, there appears to be little systematic effort along these lines, and the assignment of functions to individuals in the more complex human societies seems, at least in part, dependent on accidents. It appears that a better understanding of the genetic basis of human behavioral polymorphism would be important for a rational division of labor in human societies.

INDEX